T0248659

Making the Link

Papers commissioned as part of a study conducted by the International Service for National Agricultural Research (ISNAR)

Making the Link

Agricultural Research and Technology Transfer in Developing Countries

EDITED BY

David Kaimowitz

**Published in cooperation with the
International Service for National Agricultural Research**

Routledge
Taylor & Francis Group

LONDON AND NEW YORK

First published 1990 by Westview Press

Published 2018 by Routledge
52 Vanderbilt Avenue, New York, NY 10017
2 Park Square, Milton Park, Abingdon, Oxon OX14 4RN

Routledge is an imprint of the Taylor & Francis Group, an informa business

Copyright © 1990 by the International Service for National Agricultural Research

All rights reserved. No part of this book may be reprinted or reproduced or utilised in any form or by any electronic, mechanical, or other means, now known or hereafter invented, including photocopying and recording, or in any information storage or retrieval system, without permission in writing from the publishers.

Notice:
Product or corporate names may be trademarks or registered trademarks, and are used only for identification and explanation without intent to infringe.

Library of Congress Cataloging-in-Publication Data
Making the link : agricultural research and technology transfer in
    developing countries / edited by David Kaimowitz.
        p.   cm.—(Westview special studies in agriculture science
and policy)
    Includes bibliographical references.
    ISBN 0-8133-7896-6
    1. Agriculture—Developing countries—Technology transfer.
2. Agriculture—Research—Developing countries.   3. Agricultural
extension work—Developing countries.   I. Kaimowitz, David.
II. Series: Westview special studies in agriculture science and
policy.
S494.5.I5M346   1990
338.1'09172'4—dc20   89-48098
                                                                CIP

ISBN 13: 978-0-367-01455-1 (hbk)
ISBN 13: 978-0-367-16442-3 (pbk)

Contents

Members of the Study Group on the Links between Agricultural Research and Technology Transfer

Advisory Committee

John Coulter
David Leonard
Niels Röling

Burton Swanson
Eduardo Trigo
Taiwo Williams

ISNAR Working Group on Linkages

T. Ajibola Taylor
N'Guetta Bosso
Robin Bourgeois
Hunt Hobbs
David Kaimowitz

Deborah Merrill-Sands
Willem Stoop
Anna Wuyts
Larry Zuidema

Case Study Researchers

Dolores Alcobar, Philippines
Luis Alfonso Agudelo, Colombia
Assemien Aman, Côte d'Ivoire
Corazón Asucena, Philippines
Emiliana Bernardo, Philippines
Alexander Coles, Costa Rica
Johnson Ekpere, Nigeria
Thomas Eponou, Côte d'Ivoire
Hermina Francisco, Philippines

Eduardo Indarte, Dominican
  Republic
Ildefons Lupanga, Tanzania
Viviana Palmieri, Costa Rica
Agapito Pérez Luna, Dominican
  Republic
Kouadio Tano, Côte d'Ivoire
Soumalia Traore, Côte d'Ivoire
Germán Urrego, Colombia

Theme Paper Authors

Paul Bennell
Ruben Echeverría
Paul Engel
Peter Ewell
David Kaimowitz
David Leonard

Niels Röling
Roberto Martínez Nogueira
Carl Pray
Holly Sims
Monteze Snyder

About the Contributors

DAVID KAIMOWITZ is a research fellow at ISNAR, where he is coordinating the current international study of the links between agricultural research and technology transfer. Prior to this, he was an economist in the Ministry of Agriculture in Nicaragua. Kaimowitz has written extensively on technological change in agriculture and on rural development and agrarian reform in Nicaragua. His writing has appeared in a number of books, as well as in many journals, including *World Development*, the *Journal of Development Studies*, the *Journal of Peasant Studies*, the *IDS Bulletin*, *Public Administration and Development*, and *Development and Change*. He obtained his Ph.D in agricultural economics from the University of Wisconsin.

PAUL BENNELL has recently taken up an appointment as Professor of Economics at the University of Zimbabwe in Harare. He was formerly a research fellow at ISNAR and a research officer at the Institute of Development Studies in Sussex, Britain. Bennell has published more than a dozen refereed articles, most of which are concerned with personnel and manpower topics. He received his Ph.D in economics from the University of Sussex.

RUBEN ECHEVERRIA is a research fellow at ISNAR. His research and writing have focused mainly on public and private sector investment in agricultural research. He has worked in extension activities at the Uruguay Land Reform Institute and as a consultant to the Economics Program of the International Wheat and Maize Improvement Center (CIMMYT). Echeverría received his Ph.D in agricultural economics from the University of Minnesota.

PAUL ENGEL is Assistant Professor of Extension at the Wageningen Agricultural University in the Netherlands. He was formerly the Director of the Programme for Rural Extension in Developing Countries in the Dutch Ministry of Agriculture. Engel has also worked for the Food and Agriculture Organization (FAO) in Ghana and for Dutch Cooperation Projects in Peru and Colombia. He obtained his degree in irrigation agronomy from the Wageningen Agricultural University.

PETER EWELL is currently the regional coordinator in East Africa for the International Potato Center (CIP). His research has focused on the economics of integrated pest management. He has previously worked for the Social Science Department of CIP in Peru and as a consultant for an ISNAR study on on-farm client-oriented research (OFCOR). With T. Poleman, he co-authored *Uxpanapa: Agricultural Development in the Mexican Tropics*. His Ph.D in agricultural economics is from Cornell University.

DAVID LEONARD is Associate Professor of Political Science at the University of California, Berkeley. He has worked extensively in various capacities in Kenya and Tanzania. His books *Reaching the Peasant Farmer: Organization Theory and Practice in Kenya, Rural Administration in Kenya: A Critical Appraisal* and *Institutions of Rural Development for the Poor: Decentralization and Organizational Linkages* are well known, as are his many journal articles. His Ph.D in political science is from the University of Chicago.

ROBERTO MARTINEZ NOGUEIRA is co-Director of the Analysis Group for Institutional and Social Development (GADIS) in Buenos Aires, Argentina. Formerly, he was Undersecretary for Development Planning in Argentina. He has also worked at ISNAR, as a part-time staff member, and as a consultant to the United Nations (UN) and to the Organization of American States (OAS). He obtained his Ph.D in public administration from Cornell University.

CARL PRAY is Associate Professor of Agricultural Economics at Rutgers University. His current research is focused mainly on private sector agricultural research and seed industries. He has worked for the Agricultural Development Council in Bangladesh, the Peace Corps in India, the Department of Agriculture and Applied Economics at the University of Minnesota and as a consultant to the United States Agency for International Development (USAID) and the World Bank. His Ph.D in economic history is from the University of Pennsylvania.

NIELS ROLING is Professor of Extension at Wageningen Agricultural University in the Netherlands. He has researched and written extensively in the field of agricultural extension, and his book *Extension Sciences Information Systems in Agricultural Development* has been published in both English and Dutch. Röling obtained his Ph.D in rural sociology at Michigan State University.

HOLLY SIMS is Assistant Professor of Political Science at Claremont College, California. Prior to this, she was a lecturer at the University of California at Berkeley. Her research and writing have focused on agricultural institutions and technological change in the Indian and Pakistani Punjab. Sims obtained a Ph.D in political science from the University of California.

MONTEZE SNYDER is Assistant Professor of Political Science at the University of Florida in Tallahassee. She has conducted research on a variety of topics in West Africa and has worked for USAID and as a consultant for ISNAR. She obtained her Ph.D in public administration from Johns Hopkins University.

Preface

Policy makers in developing countries face difficult challenges in agriculture. To help finance foreign debts and import essential goods, agricultural exports must rise. Fast-growing urban centers demand more foodstuffs and industrial raw materials from the countryside. Persistent rural poverty constrains labor productivity and foments political instability. Protectionism, technological change and stagnant demand are creating downward pressure on the world market prices of many agricultural goods. In many places there is little new land available for cultivation and no simple technological "fixes" to increase production. Exisiting production is threatened by poor water and soil management and unsound agricultural practices.

Potentially, national institutions which import, generate, adapt, validate and transfer agricultural technology can be powerful tools for meeting these challenges. It is their job to identify new opportunities, help farmers and consumers solve their current problems and develop the country's knowledge base and infrastucture regarding agricultural technology.

This potential, however, is not being fully met, partly because agricultural agencies often have poor relations with the agencies responsible for delivering technological support to farmers. This results in inadequate follow-through and a breakdown in the flow of information. Thus, research efforts are less likely to be relevant and farmers are less likely to receive the information and inputs they need.

Many sources have noted the poor links between research and technology transfer in developing countries:

> Bridging the gap between research and extension is the most serious institutional problem in developing an effective research and extension system (World Bank, 1985).

> Weak linkages between the research and extension functions were identified as constraints to using the research for 16 (out of 20) of the projects evaluated (United States Agency for International Development, 1982).

> All 12 countries (in which research projects were evaluated) had difficulties of communication between research and extension agencies (Food and Agriculture Organization, 1984).

Because of the serious consequences of this problem, in 1986 agricultural research managers from a number of countries requested that the International Service for National Agricultural Research (ISNAR) conduct a study to identify key factors which influenced the effectiveness and efficiency of links between research and technology transfer and recommend ways to improve them.

In the discussions on how to implement the study, it soon became obvious that the existing literature on the subject was largely anecdotal or prescriptive, and would not provide the necessary basis for the study. A fresh approach was needed, and ISNAR decided to ask several internationally recognized experts on the subject to write a series of papers examining the relevant issues. Six papers were written, each one approaching the problem from a different, yet complementary, perspective. We then wrote a conceptual framework which pulled together the principal hypotheses put forward in these papers and developed some areas which were not covered by the papers.

This book is the outcome of these efforts. The first paper, by Niels Röling, introduces the concept of an agricultural knowledge and information system (AKIS) and the activities which such a system must perform. Röling gives some principles for managing an AKIS, including the need for user control, the importance of calibrating the different elements of the system, and the potential for linkage mechanisms to fill in the gaps which exist between these elements. He then discusses various methodologies for researching an AKIS and at the end of the paper provides a checklist of common AKIS disorders.

The authors of the second paper, Holly Sims and David Leonard, show how external pressure on research and technology transfer institutions can improve system performance. They give particular attention to the opportunities for, and limitations of, pressure from national policy makers, donors, farmers' organizations and commercial firms. Although the paper applies to a broad range of developing countries, the examples and authors' own experiences come largely from former British colonies in Asia and Africa.

The paper by Roberto Martínez Nogueira traces the growth and increasing complexity of the demands placed on research and extension agencies by policy makers in Latin America and their effect on links between the two groups. Research and extension evolved from being small organizations, which were limited in scope and personally managed by their directors, to forming part of large complex bureaucracies controlled by elaborate planning mechanisms, with multiple audiences, and an increasingly sophisticated technical division of labor. This process continued until the situation became unmanageable. Now there is a trend towards decentralization, privatization, more qualitative planning methods, horizontal coordination between government agencies, and greater integration between extension and adaptive research tasks.

Paul Bennell examines the linkage problem from the perspective of social psychology. He notes that researchers and extension agents are separate groups, each with its own background, training, experience, responsibilities, status and physical location. Occupational theories and theories about intergroup relations, including Realistic Conflict Theory, Social Identity Theory, group characteristics and intergroup contact theories, are each shown to have something to offer in helping to understand and improve the relations between these groups.

Two papers discuss important special cases of research-technology transfer links: farming systems research and the private sector. Peter Ewell looks at how on-farm, client-oriented research initiatives in nine countries of Africa, Asia and Latin America coordinated their activities with extension. Although these programs made research more relevant, they could only complement, not substitute for, specialized technology transfer efforts. Just because research is participatory, conducted on-farm and uses a systems approach does not necessarily mean there will be good relations with extension. Good relations are found only in countries that have made a strong and explicit effort to create them.

Carl Pray and Ruben Echeverría focus on research-technology transfer links within the private sector and between the private and public sectors, and on the lessons public sector managers can learn from their private counterparts. Unlike public extension, marketing is a high status activity in the private sector, with at least as much status as research. This helps to ensure research relevance and to eliminate poorly conceived research projects at an early stage. The private sector spends a greater portion of its budget on linkage activities such as preparation of promotional materials and training of marketing staff. Company size and the pattern of industrial organization in particular products heavily influence the types of links which emerge.

David Kaimowitz, Monteze Snyder and Paul Engel summarize the key points from the previous papers, grouping these points according to whether they are concerned with political, technical or organizational factors. They also add certain new elements, such as how links vary depending on the type of technology involved, the problems posed by unfamiliar environments or farming systems, and the importance of whether institutional responsibilities are divided based on the activities involved or on the type of clients.

Through reading each other's papers and engaging in ongoing discussions, many of the authors came to share similar views. Nevertheless, no attempt was made to resolve discrepancies between authors, apart from standardizing terminology as far as possible.

ISNAR is currently completing the second stage of its linkage study, involving empirical case studies in Colombia, Costa Rica, Côte d'Ivoire, the Dominican Republic, Nigeria, the Philippines and Tanzania, from which additional lessons will undoubtedly emerge. This book does, however, reflect the progress we have made on the linkage issue to date. We hope that by sharing our ideas with a wider audience we may stimulate further debate.

Funding for the research came from the Governments of Italy and the Federal Republic of Germany, the Rockefeller Foundation and ISNAR. Without their support this effort would not have been possible. Needless to say, any mistakes in the papers are our own.

David Kaimowitz
ISNAR, The Hague

1

The Agricultural
Research-Technology Transfer Interface:
A Knowledge Systems Perspective

Niels Röling

The links between agricultural research and technology transfer in developing countries are generally recognized as a major bottleneck in agricultural technology systems and have received inadequate attention in the past (Sands, 1988). A basic concept in this paper is that research and extension should not be seen as separate institutions which must somehow be linked. Instead, scientists involved in basic, strategic, applied and adaptive research, together with subject-matter specialists, village-level extension workers and farmers, should be seen as participants in a single Agricultural Knowledge and Information System (AKIS).

The concept of an AKIS has been extensively discussed in the literature, using a number of different nomenclatures and definitions (Bunting, 1986; Engel, 1987; Lionberger and Chang, 1970; Nagel, 1980; Rogers et al., 1976; Röling, 1986a and 1988a; Swanson and Claar, 1983). I define an AKIS as follows:

> An AKIS is a set of agricultural organizations and/or persons, and the links and interactions between them, engaged in such processes as the generation, transformation, transmission, storage, retrieval, integration, diffusion and utilization of knowledge and information, with the purpose of working synergically to support decision making, problem solving and innovation in a given country's agriculture or a domain thereof.

The concept of an AKIS should be distinguished from that of a management information system. The former is the entire system that produces the knowledge used in agriculture. The latter evaluates the productivity or other aspects of an enterprise (not necessarily an agricultural one) in order to help management make decisions. A substantial body of knowledge has been built up in recent years on methods for analyzing the effectiveness of management information systems, but no comparable set of methods exists for analyzing AKIS.

The Interface between Research and Technology Transfer

The AKIS serves as a conceptual framework within which to consider the interface between research and technology transfer, the central area of concern of this paper. Present knowledge about this interface is scanty. What is more, traditional methods of data gathering and analysis do not lend themselves easily to empirical research on this subject: "The lack of suitable measures to objectively assess the strength (of links between research and extension) continues to be a problem" (Sands, 1988).

Within an AKIS, the research-technology transfer interface is an especially important one in determining the performance of the whole system. The AKIS is vulnerable at this interface because major transformations of knowledge, information and technology have to take place there and because bottlenecks in their flow have grave consequences. All too often, the interface suffers from both an institutional and a functional vacuum.

Historically, research has stopped too early in what should be a continuous and dynamic process of developing and diffusing new technology. Researchers have been physically and mentally isolated from farmers and have handed down an unfinished, untested product to extension staff. Extension contact staff — squeezed between the farmers they live among, who often ridicule the technologies they bring, and their superiors, who demand results in line with policy directives — have been caught in a crisis of morale (Collinson, 1985).

Improving System Design

The total impact of an AKIS should be more than the sum of the impacts of its constituent parts: an important goal of the analysis, design and management of an AKIS is to increase the synergy of its components. Research results that remain unused, farmers without access to technology transfer services, extension that has no links with research — all are signs of an AKIS that is not operating synergically. And all provide good reasons for taking the AKIS as a whole, rather than its individual parts, into consideration when seeking to improve matters.

Recent approaches to improving agricultural technology systems, such as farming systems research (FSR) and the Training and Visit (T and V) system of extension, are efforts to improve AKIS synergy. The T and V system seeks to create regular information flows between research stations, subject-matter specialists, extension workers, contact farmers and followers. It can be seen as a management tool for improving the interconnectedness of AKIS components. FSR is a participative method for developing technology. It seeks to ensure goodness-of-fit between technology and its users, by emphasizing the importance of collecting information from and about farmers before designing technology and while testing it. FSR represents an important step toward user control as an essential ingredient of successful technology development.

On-farm research, considered by some to be a phase of FSR and by others as a desirable component of all agricultural Research and Development (R & D), allows direct contact between farmers and researchers, and has been shown to be as important an influence on applied research programs as the published results of basic and strategic research (Biggs, 1983). On-farm research is, again, a way of improving the interconnectedness of the AKIS, and is a critical step toward user control.

Even the formal incorporation of T and V, FSR or other mechanisms within an AKIS does not, however, guarantee that effective links will be established. For example, the introduction of T and V in Sri Lanka led to the formal institutionalization of a regular meeting between research and extension staff to ensure their linkage, but the following views, expressed by research officers, suggest that this dialogue between research and extension leaves something to be desired (Blok and Seegers, 1988):

In rice, mostly the problem is that extension says a recommendation is not working. Research officers then have to demonstrate that this is because the cultural practices were not followed correctly.

With respect to paddy, most problems were solved before and extension people have to be reminded of the old solution.

The chairman asked extension (workers) to be short on the reading of messages because mostly ... the research officers have heard these for seasons so they won't have much response. Even if, for example, the scientific approach to land preparation is taken up in the bi-weekly training, the research officers won't discuss why farmers didn't take it up for they already know the reason: lack of money.

Research will inform extension if there is any alteration in the recommendation.

I believe that the rapid emergence of knowledge management as an important issue for food security, agricultural sustainability and other challenges makes it imperative that we seek new tools with which to manage knowledge and information. Modeling the AKIS may provide us with such tools. Although no universally appropriate model can be developed, modeling will allow us to discern what principles of knowledge management need to be applied under given circumstances to obtain different development objectives.

This paper aims to explore the usefulness of the AKIS concept for improving the management of links between research and technology transfer. The first section looks at some of the development issues facing AKIS management. The second section discusses the components needed when modeling the AKIS. In the third section, issues in the management of AKIS are explored. The final section summarizes and draws some conclusions, focusing on knowledge management.

Management issues are the major concern of this paper. Although it is notoriously difficult to transfer successful management solutions from one context to another, it may be possible to identify some general principles. Because of the uncertainty I must attach to them at this stage, I call these principles *management hypotheses*. They appear in italics in the text of this paper.

Development Issues

This section examines some of the development issues which management must consider when designing a new AKIS, or manipulating an existing one. International interest in the links between research and technology transfer, and in the removal of bottlenecks in AKIS, reflect the fact that agriculture is becoming increasingly driven by technology. Other issues, too, have stimulated international awareness of the need for more effective AKIS management. The sustainability of agriculture is perhaps the major one, but gender and equity issues are also very important. Information technology has stimulated increased interest in AKIS management because of its potential for increasing the efficiency with which knowledge and information are created and shared.

Technology-Driven Development

Agricultural development requires a mix of conditions (Haverkort and Röling, 1984). Although the precise nature of the mix depends on the context, it usually includes good infrastructure, access to credit, water, land, markets, input delivery, social organization, relevant technology and rewarding prices (Mosher, 1966). As agriculture develops, the need for this mix is increasingly met, giving farmers more control over their environment. The greater their control, the more important knowledge and technology become as the major determinants of development (Jiggins, 1988). In other words, technology development increasingly drives agricultural development, as the other essential conditions are effectively provided for.

Even in areas where the essential conditions for development have not been met, the effect of technology-driven development is felt through competitive market pressures. New technology is usually adopted slowly at first. Early innovators (Rogers, 1983) capture large profits because they are too few to affect the market price. However, soon others copy their example. Total production begins to rise, exerting downward pressure on prices. Further down the line, farmers have to adapt in order to stay in business, but their investment already has a low return. At the end of the line, farmers see their incomes drop and can do little about it. Those who are unwilling to adapt, whose farms are too small or for whom the new technology is not appropriate, lose out. The European Economic Community (EEC) has lost some 60% of its farmers in the past 20 years through this process.

Thus improved agricultural output and efficiency result from a continuous flow of new technology, leading to reduced farm gate prices and hence to market pressures on producers to innovate in order to stay in the game. Farmers who cannot keep up are eventually squeezed out. This process is in full swing in many developing countries, or at least in some of their better-endowed areas (Lipton and Longhurst, 1985). The real incomes of the majority, especially of resource-poor farmers, in such countries as the Philippines (Cordova et al., 1981), India (Swaminathan, pers. comm.) and Sri Lanka (Wijeratne, 1988), have declined drastically in recent years.

Those leaving farming are often absorbed by the industrial and service sectors. However, non-farm employment opportunities are not growing fast enough in many areas affected by technology-driven development. It has been argued that, in such situations, AKIS management should concentrate on targeting technology development to support the livelihoods of resource-poor farmers, and on human resource development programs to help such farmers become effective users of technological opportunities (Röling, 1986b).

Improving the efficiency of an AKIS increases the degree to which technology drives agricultural development. Steering that development in desired directions, instead of allowing it to be governed only by the market, is a demanding challenge. To meet this challenge, developing countries may have to do things industrial countries have not succeeded in doing, but with fewer resources.

Benefits

If — a big "if" in most countries — alternative employment is available, the squeeze on less efficient farmers is one of the benefits of technology-driven development. When a large proportion of a country's working population is engaged in subsistence farming, the surplus labor and capital required for industrial development are scarce. By becoming more efficient, farmers free these vital resources and create the capacity to feed people doing non-agricultural jobs.

Another benefit of technology-driven development is that it enhances competitiveness. Technology-driven development implies that one must stay ahead of the game if one is to stay in business. This holds true for countries and regions, as well as for individual farmers. A country which allows its agricultural industry to fall behind must expect relatively cheap imports to squeeze out local farming, unless it raises protective barriers; a country which wants to continue exporting agricultural products must keep its competitive edge.

The need to compete is the major argument advanced by governments to support their expenditure on agricultural research and extension. The task of research is to provide a steady stream of new technologies to agriculture, and especially to those sectors producing export crops. The technology stream — note the "one-way" metaphor — is seen as the product of a sophisticated AKIS dedicated to technology-driven development. Badly functioning links between research and extension disrupt the flow of the stream or, at worst, prevent it from flowing at all.

A third benefit of technology-driven development is that, where it occurs, the promotion of new agricultural technology requires relatively little attention. In agriculture, a relatively small effort can lead to the rapid diffusion of new technology, as the price mechanism makes the process compulsive. Because early innovators make the largest profits, these so-called "progressive" farmers become so keen for new ideas that they exert strong pressure on the AKIS to provide more of them. Few industries are so innovative as agriculture. All a government has to do is to provide an efficient AKIS to feed new technology to the progressive farmers, who are relatively few in number (about 20% of the total), and continue to satisfy the needs of the agricultural development mix. Market forces do the rest.

Once technology-driven development takes off, the AKIS manager can rely on market forces to accelerate diffusion and to squeeze out inefficient and resource-poor farmers, if that is considered desirable.

These are the benefits of technology-driven development. However, there are important drawbacks for national policy makers selecting this development path. These drawbacks concern rural livelihoods, rainfed areas and sustainable agriculture.

Drawbacks

Rural livelihoods. As technology-driven development occurs, many developing countries find it impossible to expand alternative employment fast enough to accommodate those leaving the land. Moreover, their rapidly growing rural populations increase the pressure on land, reducing farm sizes with each new generation (Safilios Rothchild and Mburugu, 1987; Swaminathan, pers. comm.).

In Asia, 55% of farmers now cultivate an area of less than 1 ha each and 73% farm less than 2 ha. In the world, some 280 million farmers cultivate less than 1 ha each (World Bank, 1982). Assuming an average farm family of six, that represents some 1.5 billion people. These people seem prime targets for technology-driven marginalization rather than development, since alternative employment for such numbers seems out of the question in the foreseeable future.

Indeed, a rapid deterioration of the livelihoods of many rural families, especially in non-irrigated areas, is already occurring. Millions of farmers in such areas face falling farm gate prices and reduced income-earning opportunities. Zimbabwe provides an example.

Small-scale farmers in Zimbabwe have made tremendous strides in adopting modern maize technology. In the high-potential communal areas of Manicaland, in the eastern part of the country, 1 acre of traditional maize usually yielded only about four bags of grain. Concerted extension efforts over a number of years have raised production to 20 bags per acre. At 1986 prices, 10 bags of grain were enough to feed an average farm family, while another 10 were needed to offset the purchase cost of inputs, such as hybrid seed and fertilizer. However, since 1986 the rising costs of inputs, combined with the lower maize price now set by government (under pressure, because of unsold stocks), have raised the number of bags at which a farmer can break even. Farmers using inputs may suffer heavy financial as well as crop losses in the event of drought. The falling maize price soon forces small-scale farmers to expand the area cultivated. Increased operations require tractor-aided land preparation, lorry transport of inputs and harvests, seasonal credit, and so on. Eventually, unless a cheaper technology can be provided, only the large-scale mechanized farmers will be able to stay in business.

The case of Zimbabwe, a typical victim of its own success, illustrates the human costs of technology-driven development. Even if global food surpluses are short-lived, severe disruption seems likely, with many people losing their livelihoods and being forced to leave the land. At present, about one third of urban growth in developing countries can be attributed to rural-urban migration (Bronsema and van Nimwegen, 1987), although of course not all of this is caused by surplus food production.

The fact that food has become less expensive does not solve the livelihood issue for those who have diminishing purchasing power. Seen in this light, agricultural development is no longer a means to end rural hunger and scarcity. What it does under conditions of surplus is to reallocate resources to areas and people who can produce most efficiently. Agricultural development, in the absence of countervailing measures, has high costs in terms of the number of people whose livelihoods are adversely affected.

Hence there is a need to create an AKIS capable of performing roles other than the promotion of technology-driven development. So far, AKIS management in most countries has emphasized the provision of continuous opportunities for innovation to progressive farmers (the case of Pakistan, documented by Sofranko et al. in 1988, is typical). This has driven agricultural change and brought us the luxury of surplus — but in a market limited by low purchasing power. In future, the development and dissemination of technology will need stronger controls and more deliberate targeting toward small-scale producers, with an emphasis on low-cost, scale-neutral technology.

Such an approach will make heavy demands on the interface between research and technology transfer. The extreme heterogeneity of small-scale producers will require a substantial increase in the resources devoted to technology adaptation. The need for more careful targeting implies a greater flow of information about farmers to research, and more emphasis on mechanisms for enhancing user control of the technology development process.

Rainfed areas. International and national AKIS have in the past focused largely on the relatively easy task of serving irrigated cropping systems (Castillo, 1983). In developing countries, about one fifth of cultivated land is now irrigated. This land area uses about 60% of all fertilizers and produces 40% of all annual crops in the developing world. Between 50 and 60% of the increase in agricultural output in the past 20 years has come from new or rehabilitated irrigated areas. However, such areas support only about one third of the world's farmers (World Bank, 1982), and are not very important in Africa.

Irrigated areas, and a few high-potential rainfed areas, are characterized by a high degree of farmer control over a homogeneous environment. Providing a steady flow of appropriate technology for the vast majority of rainfed areas in which conditions are far more variable has not so far been achieved. Yet most arable land and most of the world's farmers depend on rainfall that is often highly erratic.

For a variety of reasons, including increasing problems of silting, soil salinity, pest resistance to chemicals and the pollution of water resources with nitrates, crop yields in irrigated areas have reached a ceiling in recent years. And the potential for further increasing production in what have hitherto been known as the high-potential rainfed areas may likewise be limited for environmental reasons. This means that the world's next quantum leap in food production will probably have to come from what are currently considered to be medium- to low-potential areas of rainfed agriculture.

Serving heterogenous agricultural systems has considerable implications for an AKIS. The strong performance achieved through disciplinary specialization and a focus on few commodities must be traded off against the need to ensure that technology fits complicated agricultural systems with a wide range of commodities and farmer activities. The case of Turkey illustrates the need to develop more flexible AKIS models under such circumstances.

Agriculture in central Turkey is highly diverse. There are mobile beekeepers who move their hives on lorries from one flowering crop to another; there is extensive wheat production under a virtually feudal system; there are also small-scale village farmers growing wheat, vegetables and various fruit and nut trees. Such farmers often practise small-scale irrigation along streams and rivers. Large numbers of cattle, sheep and goats are kept by both village farmers and semi-nomadic herders. Some producers are investing in new enterprises such as broiler production, sunflowers or sour cherries, and new innovations, including sprinkler irrigation.

To serve this rapid diversification, the Government of Turkey is experimenting with the T and V system in 16 provinces. Subject-matter specialists regularly train extension staff; these specialists are, in turn, trained by researchers. However, the typical province has a commodity research institute. As a result the T and V system operates somewhat irrationally: extension workers trained by a subject-matter specialist in wheat may be asked to help farmers with their vegetable production, while the local commodity institute responsible for providing technical backstopping specializes in sunflowers.

In Turkey as elsewhere, rainfall variability complicates matters still further. Links between research and technology transfer must respond to seasonal or inter-annual fluctuations. Producers' knowledge of local conditions assumes a high value.

If farmers in high-potential areas continue to improve their efficiency, they will drive agricultural prices below the level at which it is feasible for small-scale producers in medium-potential rainfed areas to use modern inputs and varieties. This implies that farming in such areas will require a technology based on different principles (Haverkort, 1988). And even if production in high-potential areas stagnates, so that world food prices begin to rise again, the green revolution approach to technology development will still prove largely inapplicable throughout vast areas of medium-to low-potential land in which erratic rainfall makes the use of purchased inputs too risky.

Some rainfed areas will not be able to generate enough marketable surplus to justify expensive research stations, subject-matter specialists, facilities for delivering inputs, and all the other services we have come to associate with a modern AKIS. In such circumstances the design of an AKIS will need to focus mainly on farmer experimentation, local knowledge, and local networks for promoting the use of technology (Chambers and Jiggins, 1987; de Vries, in prep.).

Sustainable agriculture. We live in a world in which every day brings new signs that our environment is deteriorating. The major threats are the greenhouse effect, increasing soil salinity and erosion, flooding, the growing resistance of pests and diseases to chemicals, deforestation and desertification, and surface and ground water pollution and depletion. Many so-called "traditional" agricultural systems have, in fact, become systems undergoing rapid degradation.

The shifting cultivation system of West and East Africa provides an example (Bantje, 1987). The first sign of trouble is a shorter fallow period. A typical 1-acre family farm requires 24 acres of land for a 23-year fallow period, assuming 1 year of cultivation. Dividing up this land between three sons in equal blocks of 8 acres reduces the fallow period to 7 years. Further dividing each block of 8 acres between two grandsons reduces it to 3 years, a period inadequate to restore soil fertility, even without taking into account the added pressures resulting from grazing and the demand for firewood.

Reduced soil fertility requires larger fields and multiple use of the same field to produce the same quantity of food. If additional production for cash is required, this only aggravates the problem. Farmers start to use purchased fertilizers. These are increasingly expensive, however, and do not pay for themselves when the produce is used for subsistence. The land continues to be divided up. The next step is the widespread use of cassava, a crop that mines the soil, as the main staple crop, replacing millet, sorghum and other traditional cereals. Throughout the process, men increasingly migrate to find urban wage employment, leaving their wives and children on the land.

The development of more sustainable agricultural systems has become a high priority for many governments. This requires a different kind of AKIS to the model used in development that is driven purely by technology. An AKIS dedicated to sustainable agriculture would place far more emphasis on assessing the likely impact of research and technology transfer activities before carrying out such activities on a large scale (International Rice Research Institute, 1987; World Commission on Environment and Development, 1987). In addition, the activities themselves would be different— and they would be implemented differently. Soil conservation, for example, requires a much more persuasive approach than the promotion of hybrid maize, and often needs to be supported by policy measures. This, in turn, requires sophisticated links between research, technology transfer and policy bodies.

Conclusions

Efficiency and competition, on the one hand, and sustainable agriculture and livelihoods on the other, are uneasy bedfellows. Focus on the former reinforces technology-driven development, leading to the preponderance of high-tech solutions, to the squeezing out of inefficient producers, and to a tendency for short-term economic interests to overrule long-term ecological considerations. Sustaining livelihoods and environments is a different process altogether. To the extent that individual countries can still control their own national agricultural development, a policy focus on sustainability requires instruments other than a reliance on market forces.

Modeling the System

Basic Concepts

The study of knowledge. The dimensions along which knowledge evolves are not easy to specify. The claim that its evolution follows a dimension of increasing control over the physical environment seems defensible (Ascroft, 1972; Röling, 1970). If adaptation and control are two ends of a continuum, the evolution of knowledge can be said to allow people increasingly to control instead of adapt. But this evolution has now brought us to the point where we meet ourselves: the environment is largely man-made and survival implies voluntary changes in human behavior for the common and future good. In other words, it is now ourselves, not the environment, that we need to control. It is as yet uncertain whether this will be feasible in time and on a sufficient scale, but the use to which knowledge is put will probably determine whether or not we survive in this new phase of human existence.

Given the fundamental role of knowledge in human survival, it is surprising that the study of knowledge and its use has taken such a long time to emerge. However, knowledge processes such as diffusion (for example, Rogers, 1983) and utilization (for example, Beal et al., 1986; Havelock, 1969) have received increasing attention over the past 25 years. The study of agriculture provides a dramatic case of knowledge processes in action (Lionberger, 1986).

Knowledge, information and technology. Before we proceed, some crucial terminology must be clarified.

Knowledge is a property of the mind, and cannot be transmitted to others unless transformed or encoded. Knowledge processes (memory storage, transformation, etc) are intra-personal.

Information, on the other hand, can be transmitted to others. It consists of a pattern imposed on data which simultaneously affects the interpretation of those data and enables them to be transmitted. It has been defined as a "difference in matter-energy which affects uncertainty in a situation where choice exists among a set of alternatives" (Rogers and Kincaid, 1981).

The difference between knowledge and information is one that is recognized by communication theorists. Meanings are in people. They can be encoded in messages by the source, but these messages must in turn be decoded by the receiver (Berlo, 1960). In the process, two transformations take place, making the direct transfer of what the source had in mind unlikely. Information is thus a relative concept, depending on the uncertainty experienced by its receiver. Transmitting information is a risky business because one never knows what will be informative to the intended receiver.

Technology is the software and hardware available for controlling the environment for human purposes. The software consists of methods and skills. The hardware consists of physical objects, such as tools, equipment and genetic material. Technology development can be based on the advance of science and hence on the application of research findings. However, these are by no means the only sources of technology development. Technology is continuously developed on farms, in kitchens, in backyards or in shops, by farmers, cooks, tinkers, tailors and so on. Whatever form it may take, technology is the means by which inputs are transformed into outputs (Fresco, 1986).

Definitions of an AKIS. An AKIS can be defined in three different ways: (1) as sets of organizations and people engaged in knowledge and information processes (as defined in the introductory paragraphs of this paper); (2) as sets of coherent cognitions that have evolved among members of organizations, communities or societies; and (3) as computer-based "intelligent" software (for example, expert systems, artificial intelligence). I will briefly discuss the first two definitions.

The institutional and cognitive definitions are not, of course, independent of each other. I speak of knowledge and information systems to stress that both the cognitive processes taking place in people's minds, and the communication processes taking place between them and within/between institutions, are essential for understanding an AKIS. Yet there is a difference in emphasis between those who take the cognitive view and those who, like myself, emphasize institutions and the links between them.

When an AKIS is seen as a cognitive system, the components of the system are cognitions, that is, concepts, theories and beliefs about "reality" that guide our behavior (Röling, 1986c). The cognitive approach, based on ideas in the literature on the sociology of knowledge and cognitive anthropology, has been further developed by various authors at the Department of Development Sociology, University of Wageningen, including Arce and Long (1987) and van der Ploeg (1987), who have emphasized its use in comparing the perceptions of farmers with those of other participants in the AKIS in order to reveal their different assumptions and expectations. They argue, rightly, that understanding how farmers make decisions in such areas as the recruitment of labor or the mix of enterprises on the farm are vital for the development of appropriate new technology.

The cognitive approach has been used to explore several aspects of reality as perceived by the farming family, including the classification of weeds, and male/female uses of cassava (Jiggins, 1986). My favorite example of the application of this approach is the "image of the limited good", a concept coined by Foster (1976) which I applied to fertilizer use in Nigeria.

I was working in Nigeria at a time (1966) when the Food and Agriculture Organization (FAO) was starting up its fertilizer program there. The manager of the program told me that chiefs in a number of villages had asked him to remove the fertilizers because they caused considerable discord among villagers. Apparently, people believed fertilizers were a "medicine" which could pull fertility from neighboring fields to the one in which the fertilizer was applied. The underlying assumption was that fertility is a "limited good", so that if one person gets more of it, another must necessarily get less. We laughed at the time. Now, we realize that the villagers were probably right after all, though at a different level of the system than the village.

Instead of investigating cognitions, the institutional approach looks at sets of interconnected actors, each engaged in different activities, such as research, technology transfer, production or consumption, and each playing different but complementary, roles and hence functioning synergically. Havelock (1986) provides an example of the institutional viewpoint in use in his description of the US land grant system:

> The oldest, most elaborate, most ambitious, and arguably most successful effort to develop a structured macrosystem for knowledge development and use has been going on in agriculture in the United States over the last 100 years. The land grant universities, their

experiment stations, and the Cooperative Extension Service together comprise a coherent and well-coordinated system for the generation, transmission, transformation and utilization of scientific knowledge about agriculture, home economics, and to some extent community development and youth development. These truly amazing feats of knowledge production and use are realized through an elaborate sequence of institutions and mechanisms, partly consecutive in mission, partly redundant. It includes some unique institutions and roles such as the extension specialist and the county agent, people who act as linkers or boundary spanners between the worlds of research and development, and the world of routine everyday practice.

The institutional approach leads to theory building about the way people and organizations receive, transform and transmit information, about the interfaces between them, and about the complementary roles institutions play in relation to each other. The purpose of the approach is to improve the management or design of the AKIS so as to make it function in the ways deemed desirable by policy makers, farmers and other participants in the system. It is my contention that, if only the actors in an AKIS would begin to see themselves and other actors as playing complementary roles, many AKIS would "auto-improve".

Basic Processes

In the definition of an AKIS with which I introduced this paper, I named a number of basic knowledge processes, including generation, transformation, integration, storage and retrieval. I will now discuss these processes. In so doing I will show how all the actors in an AKIS are engaged in all its basic processes (Engel, 1987).

All participants in an AKIS engage in all its basic processes.

Generation. This process is typically attributed only to research. Yet public agricultural research is not more than 100 years old in most countries. Farmers have managed for thousands of years to develop their agriculture. Many, if not most, of the improvements in agricultural knowledge today are based on their work. Farmers are, in fact, researchers. I am well aware that the term "users" does not do sufficient justice to farmers as researchers.

Many types of farming are even now not served by public research. Examples include organic farming in Europe (de Vries, in prep.) and many of the rainfed production systems on which two thirds of the world's farmers depend. It is often argued that these systems will never generate enough surplus to warrant setting up an elaborate AKIS. This is a dangerous argument, since it presupposes an unsuccessful outcome to research. However, there are undoubtedly some cases in which the costs of sophisticated research so far outweigh its potential benefits as to make it uneconomic for the time being. Until the economic picture changes, efforts in such cases should focus on teaching farmers to carry out their own research more systematically (Chambers and Jiggins, 1987; Haverkort, 1988).

Knowledge generation appears to be more effective when carried out in groups than when attempted by individuals. Empirical studies have shown that the productivity of researchers is related to the extent to which they participate in networks.

Transformation. The transformation of knowledge is perhaps the most crucial process taking place in the AKIS. The essence of an AKIS is that knowledge generated in one part of the system is turned into information for use in another part of the system. This transformation process is not well understood. Elsewhere (Röling, 1988a), I have suggested that the transformations taking place within an AKIS are as follows:

1. From information on local farming systems to research problems
2. From research problems to research findings
3. From research findings to tentative solutions to problems (technologies)
4. From technologies to prototype recommendations for testing in farmers' fields
5. From recommendations to observations of farmer behavior (male, female, children)
6. From technical recommendations to information affecting service (inputs and marketing) behavior
7. From adapted recommendations to information dissemination by extension
8. From extension information to farmer knowledge

Training people in how to transform knowledge in their part of the AKIS is a prerequisite for a more effective AKIS. Often farmers unwittingly apply a scientific principle without knowing why it works and without being able, therefore, to improve the practice (Blum, pers. comm.). Few extension workers are trained to transform a technical recommendation into instructions a farmer can follow, or to assess the demands a recommendation makes on a farmer's resources.

Integration. Like the other basic processes, integration is carried out by all participants in an AKIS. The review articles produced by scientific disciplines to pull together research results are obvious examples. The leaders of multidisciplinary research teams are engaged in a continuous effort to integrate the research results produced by different disciplines. However, little is known about how farmers integrate knowledge and information.

We assume that research results are simply turned into technologies for farmers' use but the reality is more complicated. Often, pieces of information that have been available in different fields for some time are combined to form useful new ideas. Farmers must integrate external information from many sources — other farmers, specialist literature, scientists, technology transfer services, etc — with their knowledge of their own circumstances.

Integration is often hampered by the absence of area-specific experiment stations, of FSR teams capable of intergrating disciplinary knowledge and adapting it to system needs, and of subject-matter specialists in extension services. However, examples of successful integration can be found, as in the case of mango production in Queensland, Australia.

The mango industry developed rapidly in tropical Queensland in response to good markets in major cities in Australia and abroad. The example of a few highly successful pioneers led to rapid diffusion. As a result, there was considerable pressure on the Horticulture Division of the Queensland Department of Primary Industries (QDPI) to provide more information. It soon turned out that research on mango was published by different institutes in different regions and that there was no single place where information on the results had been brought together, let alone integrated. A senior extensionist of the Horticultural Marketing Extension Service at the Bowen research station was put in charge of collection and integration. All other institutions active in mango research undertook to keep the extensionist abreast of all developments (Hubbert, 1987).

Storage and retrieval. These processes would seem to be typically the task of specialized libraries, but most scientists have their own collections of materials which they access more or less satisfactorily. Extension workers and farmers also store and retrieve information. Apparently, village-level workers in Sri Lanka seldom act immediately to pass on to farmers the information they receive through the fortnightly meetings organized under the T and V extension system, because it is often not relevant at that time; instead, they store the information for use in the future. The storage device they use is, simply, memory. Few hand-outs are given to village-level workers and they do not have manuals to which to refer.

Theoretical Models

No single AKIS model can be developed for both low- and high-potential agricultural production systems, equally suitable for both highly specialized horticulture in industrialized countries and extensive nomadic herding in, for example, the Sahel. Nonetheless, two basic types of model can be discerned.

One-way models. The most common and influential model of an AKIS is the "transfer of technology" (TOT) model (Chambers, 1983). In this model, researchers play the glory role of creating "breakthroughs". These breakthroughs are "transferred" to extension for "delivery" to "users". "Scientists develop the product and extension has to sell it" (Bennett, 1988; Chambers and Jiggins, 1987). Since researchers are among the most powerful members of the AKIS, and tend to become its administrators and managers, the TOT model is hard to replace in most national and international settings.

This is unfortunate, because analysis suggests that the TOT model is not an appropriate one. The model does not reflect the experience of successful systems, examples of which are the US Land Grant model, the Dutch farm development system, the Taiwanese system, and the R & D system which has been developed by Multifulcrum, a large and well-known multinational corporation. (The case of Multifulcrum, described in further detail below, is a real one but we have changed the name of the corporation for reasons of confidentiality.)

The TOT model assumes a linear, one-way process, starting with the breakthrough at the international level and ending on the farm with an adapted innovation. It usually succeeds in delivering technology only to progressive farmers (Röling, 1988b). The model reflects inadequate understanding of the nature of knowledge systems, yet many researchers continue to adhere to it. An example is the analytical framework developed by the International Program for Agricultural Knowledge Systems (INTERPAKS) of the University of Illinois and illustrated in Figure 1 *overleaf* (Sands, 1988; Swanson, 1986).

INTERPAKS has done much to call international attention to the fact that extension, research and other AKIS components should be seen as forming part of a synergic whole. INTERPAKS has also pioneered the search for indicators of effective AKIS (Sands, 1988). The one-way nature of the INTERPAKS model is defended on the basis that, though farmer-initiated innovation is important, the essential process in agricultural modernization is science-based innovation. Hence the INTERPAKS model depicts the AKIS as a one-way, linear system. I believe that only some technical innovations are science-based; policy-driven, market-driven and farmer-driven innovations seem equally, if not more, important (Kaimowitz et al., 1989).

To sum up, models of the AKIS which do not reflect some flow of information and influence from technology users to other parts of the system are

Figure 1. The INTERPAKS framework for analyzing agricultural knowledge systems

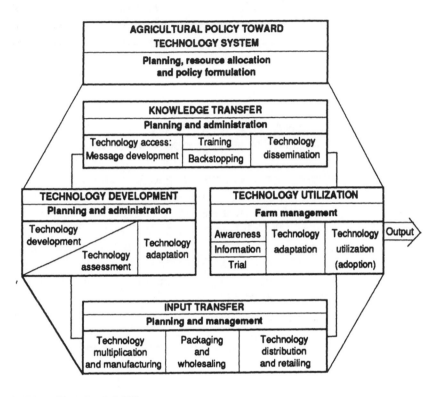

Source: Adapted from Sands (1988)

misleading. This is not because of the moral principle that participation is needed if the system is to function benignly. It is because empirical evidence from effective AKIS clearly demonstrates that user control in some form is an essential ingredient.

Two-way models. Two-way models of the AKIS are less widely used than the TOT model, but have nevertheless been developed by a number of authors, including Havelock (1969, 1986). In Havelock's model (*see* Figure 2) there is still a distinction between those who generate technology (the resource community) and those who implement it (the user community), but it is recognized that the *exchange* of information between the two communities is crucial to successful technology generation and transfer. Moreover, the resource community is not seen as the *only* source of useful information to the user community. Farmers, and not just scientists, may generate information and technology. Similarly, resource communities may become *users* of information and technology, both from farmers and from other resource communities. In other words, the roles of the two communities are seen as less stereotyped and more interchangeable. The two-way character of Havelock's model has been neglected, yet all AKIS models which take user control into account must be based on such two-way links.

Figure 2. A two-way model of the AKIS

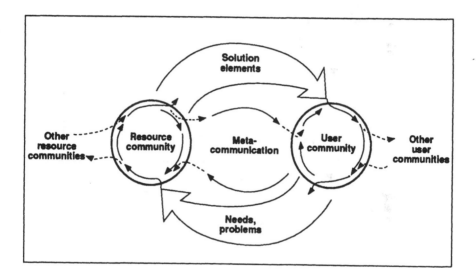

Source: Adapted from Havelock (1986)

Effective links between resource and user communities cannot be established without prior communication between the two communities — communication that establishes agendas, ground rules, appropriate media and an understanding of internal processes and contextual factors. Within the figure, all these additional communicative elements are identified as "metacommunication", that is, communication about communication.

Existing Models

Two contrasting cases. The contrasting examples of models applied in South Korea and China show not only how different models have been applied in different places and at different times, but also how each model has its own strengths and weaknesses. The strengths of a specific AKIS cannot easily be transferred to other sectors or other countries.

South Korea is said to have an effective top-down system that allows it to push technologies to farmers rapidly. Improved varieties quickly made the country self-sufficient in rice during the 1970s. But these varieties were vulnerable to cold weather and blast, both of which finally hit crops in 1979 and 1980. Ironically, the failures of those years can be attributed to the strength of the Korean guidance service which, in contrast to the extension service in most developing countries, was able to transform research into production. In the single-minded pursuit of its political goals it neglected elementary precautions that might have avoided the problems of 1979 and 1980 (Steinberg et al., 1982).

In the People's Republic of China during the 1960s and 1970s the system worked the other way round. Wuyts (1988) writes:

Probably this system's most significant aspect is the active involve-ment of the users themselves in the process of experimentation. Sheridan (1980) terms the practice "peasant innovation". The impres-sion given is that peasant innovativeness was not created by the integrated system, but that the system recognized it and adopted policies to promote and encourage it and, more significantly, to have scientists acknowledge it. Those joint efforts produced scientific farm-ers actively and enthusiastically pursuing experiments varying from different cropping patterns, testing new technology, incorporating modifications for adoptability, exchanging information with scien-tists about ...practices which could be improved and popularized (similar to the development of herbal medicine), etc.... But the system was not without its costs. One major aspect was the neglect of expertise and basic science research. This led many observers to remark (on) the poor quality of the work done as well as the low level of staff competence.

The AKIS as Part of a Larger System

When modeling the AKIS, it is important to bear in mind that the system takes its place in a larger context, from which it is not separate (*see* Figure 3). Agricultural knowledge and information processes must be examined at national level against the backdrop of: (1) the policy environment, which formulates the laws and incentives that influence agricultural performance; (2) structural conditions, such as markets, inputs, the resource base, infra-structure, and the structure of farming; (3) the political and bureaucratic structure through which interest groups influence the system; and (4) the external sector, comprising donor agencies, international agricultural research centers (IARCs) and/or commercial firms (Elliott, 1987).

Figure 3. The AKIS as part of a larger system

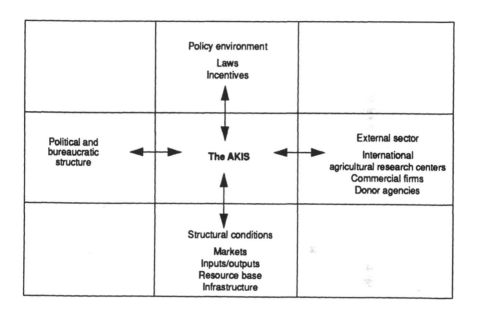

Source: Adapted from Elliott (1987)

The policy environment plays a crucial role, so much so that in some AKIS models it is considered one of the components of the AKIS itself. In this paper, I consider policy as a prime mover outside the AKIS. Together with two prime movers inside the system, namely management and user control, policy is considered as a force which can overcome the default conditions to which a system reverts unless pressures are applied to prevent it from doing so (Sims and Leonard, 1989). Deliberate policy goals and policy support are necessary, for example, if the default condition of serving progressive farmers only is to be avoided.

Likewise, structural conditions play an important role. Variability in the production environment and among the farmers who use it has tremendous implications for the design and management of the AKIS. Technology development and information transfer must be coordinated with other elements of the mix of conditions necessary for agricultural development, such as seed and input distribution, the provision of infrastructure, and so forth. This, in turn, requires horizontal links between the institutions comprising the AKIS and with other institutions outside it.

The political and bureaucratic structure also influences the performance of the AKIS. Often, the very fact that the AKIS, with its inherent need for farmer orientation and user control, is part of a bureaucratic system is enough to ensure that strong default conditions obtain. Under default conditions the bureaucracy serves its own interests and those of its political masters, rather than those of its clients. The forms of bureaucracy used and the ways in which these forms are linked, as well as the historical stage of evolution reached by the different organizations involved, are all important factors (Martínez Nogueira, 1989).

The external sector contains another prime mover (donor agencies) and acts as a source of information and technology for the AKIS itself. In some cases, for example where an IARC cooperates closely with a national research team, parts of the external sector may be considered as belonging to the AKIS.

System Research Methods

It is difficult to design research on AKIS. The traditional quantitative research methods of social science (surveys) are not applicable, although they can be used in market research to assess the potential uses of new information or technology. Despite the difficulties, I can identify a number of possible approaches to the empirical analysis of AKIS. These are:

1. *Comparative analysis:* the study of AKIS characteristics and processes in relation to system performance, comparing different AKIS in terms of their major components, linkage mechanisms, management decisions and so forth. An example is a study comparing the AKIS for cocoa with that for maize in Ghana (Annor-Frempong, 1988).
2. *Comparing formal and actual AKIS:* by comparing the official system and the actual system, and looking at the reasons for the divergence, insights can be gained into the forces operating in actual systems. Examples of this method include studies of the Sri Lankan system (Wijeratne, 1988; Blok and Seegers, 1988).
3. *Designing a matrix of major system components and filling in the cells (links):* the matrix forces one to identify all possible links and systematically investigate them. Blok and Seegers (1988) used such a matrix, as also did van Beek (1988).
4. *Following a specific innovation through the system:* working backwards from a successful innovation, one could try to trace its path. De Soyza

(1988) followed farmer problems which made it upstream to research. Over several years, only five such problems were identified, in itself a significant observation. Unfortunately, the method does not allow conclusions to be drawn as to why other problems did not make it.

5. *Using the theory as an overlay:* a theoretical model of the AKIS can be compared with what actually exists, in order to identify institutional and functional gaps.

6. *Analyzing transformations:* this would involve looking at the changes that take place when, for example, research findings are transformed into recommendations, or farmers' problems are transformed into researchable issues.

7. *Participating in events:* participating in major meetings and observing what takes place there provides much insight into AKIS operations and links. Meetings could, for example, be rated according to whether the representatives from extension raise farmers' problems and, if so, what the reactions of scientists are. Wagemans (1987), de Soyza (1988) and Wijeratne (1988) have used this method.

8. *Assessing corporate ideology:* investigating the extent to which actors in the system see themselves as playing complementary roles in the system, or as playing non-system roles.

Managing the System

No one person manages the entire AKIS. Each farmer is his or her own boss; and the numerous institutes that also make up the AKIS each have their own management team. Senior government officials may exert considerable influence over both the AKIS and the policy environment in which it operates, but their influence, while crucial in some areas, neither can nor should be all-pervasive. Much depends on the ability and willingness of government to relinquish power to other groups within the AKIS, notably the users, while retaining control, through appropriate policy instruments, over areas vital to the common interest.

The Need for User Control

Information about users should have leverage within an AKIS, not least because of the ultimate power of the farmer to refuse to use new technology. In many countries, far too little information about farmers is still being used in the technology development process. As a result, farmers refuse to innovate and agriculture stagnates. One of the most notorious examples of neglecting information about users occurred in Africa, where the role of women farmers in agriculture was for many years almost completely ignored by male-dominated AKIS bureaucracies.

Consideration of user needs is relevant not only for farmers but for all participants of the AKIS who play user roles. A technology developer has to anticipate what extension agents will accept and need, and a basic or strategic researcher has to anticipate the needs of the applied or adaptive researcher.

Participants in an effective AKIS take the needs of users into account at each point at which information or technology is transformed or adapted.

When effective AKIS are compared, user control in some form appears to be an essential ingredient. The USA and the Netherlands are respectively the first and the second largest agricultural exporters in value terms in the world. Both require a highly efficient AKIS to maintain their competitive edge. Neither of their systems can be adequately depicted as a one-way street.

Advocating user control goes against the grain of both bureaucrats and researchers, who like to think of the world as hierarchically structured, with themselves close to the top. In this respect most public sector bureaucracies contrast sharply with multinational corporations such as Multifulcrum, which builds user control deliberately into its R & D so as to ensure market orientation.

The US system. One mechanism for exerting user control in the US system is farmer influence on the county extension agents. Their salaries are paid partly by the counties so that farmers in the county can exert pressure on them (Woods, pers. comm.). It is not certain how much influence farmers have over research, however. Although he mentions user control as one of the key attributes of the US AKIS, Rogers (1986) states that "while much rhetoric is given to this feedback about needed research from farmers through the extension service to agricultural scientists, it is actually a fairly rare occurrence." Yet the political influence of the farm lobby in state legislative bodies and hence on the research budgets of state universities and land grant colleges is substantial. What is more, many of the researchers in such universities and colleges came from a farming background themselves and have experienced extension programs at first hand.

As in other countries, the farm lobby in the USA is dominated by large-scale, progressive farmers. However, these influential farmers, though by no means modal farmers, are much more representative of a sizeable portion of the farming community than the farmers who wield influence in most developing countries. Medium- and even small-scale farmers in the USA often have a similar degree of control over their environment to that achieved by large-scale farmers and found on research stations. Thus, the results of research are likely to be applicable across a broader range of farm size than in developing countries. Yet even in the USA the influence of elite farmers has been blamed for many malfunctions of the AKIS.

The Dutch system. In the Netherlands, the system is still (1989) largely government based, but it has similar characteristics to those of the US system. Many researchers and extension workers are from a farming background. Extension workers are highly service-oriented (Bos and Burgers, 1982) and virtually ignore "The Hague" and their regional supervisors. Some 50% of the costs of experiment stations and experimental farms are paid for by farmers, who also participate in program committees (although here the rhetoric overstates their degree of actual control over the research program). Specialist officers at the national level link research institutes and experiment stations with regional field work, and are responsible for the flow of information about field problems to the research community. Farmers' unions have a national apex structure which provides for the representation of farm interests on most important boards and councils at national, provincial and local level.

However, farmer influence in the Netherlands is so strong that it is difficult to control pollution caused by farmers. The pollution of air and water by surplus manure is especially problematic. The results of research linking acid rain to farm manure and providing evidence of soil pollution have been withheld, and have been acted upon only belatedly.

One of the reasons for the present plans to privatize the extension service is that it is difficult to combine farmer-controlled, client-oriented extension services with the implementation of government policy, whenever the latter is no longer aligned with farmers' interests. Separating the two will, it is hoped, resolve this conflict of interests. Privatization of technology transfer in Britain, Belgium and parts of Australia is occurring for similar reasons. An important question for these countries is how publicly funded research will link with private technology transfer.

The Multifulcrum system. Multifulcrum's R & D is based at its central research laboratories in Europe. The laboratories serve a set of international companies, which also have departments for R & D, as well as for marketing and production. These R & D departments link the companies with the central laboratories. Management at the laboratories has ensured that the R & D department of each company is headed by a person recruited from the laboratories (a linkage mechanism called "body swapping", see Improving Linkage Mechanisms).

The companies are heavily market-oriented. They carry out a great deal of market research among consumers, on topics such as consumer categories, preferences, buying decisions, and reactions to prototype products. This market information is supplemented by technical insights, obtained by a special team of researchers which visits families, observing and even video taping their handling of company products.

A complex procedure is used to decide on research projects. Research proposals are formulated by ad hoc study groups in each company, consisting of the company chairman and representatives of its marketing and R & D divisions, as well as a manager from the laboratories. These proposals are sent to a single Steering Committee that covers all the companies. The committee meets once every two years to decide on major research directions.

In the light of these decisions, R & D projects are designed and supervised by project groups. These groups consist of representatives of company R & D departments, together with laboratory staff. A Project Group Manager is appointed jointly by the Steering Committee and the management of a company. He or she formulates the project, including its strategic justification, key objectives, workplan and budget statement.

Great care is taken when such projects are approved. Each project must meet the approval not only of the R & D Steering Committee but also of the marketing, technical and R & D representatives of the company concerned. Even at this late stage the key objective might be adjusted. Each year projects approved are included in the R & D year-plan, which is compiled by the Steering Committee. An annual contract between the companies and the laboratories is drawn up and signed for the research to be carried out that year in accordance with the year-plan.

An interesting aspect of the set-up at Multifulcrum is the sharp distinction between basic and applied research. Undifferentiated "Research" with a capital "R" does not exist at Multifulcrum. Basic research of a disciplinary nature serves all the Multifulcrum group's major branches and is carefully protected from market pressures, whereas applied research is carried out independently by each product group and is strongly tied to consumers' needs.

Although basic research is shielded from market pressures, it too must be productive. The mechanism for ensuring productivity without undue interference is to make basic research financially dependent on the product groups, but to deny the latter any involvement in project reviews. The procedure followed is to allocate a certain percentage of each product group's research budget to "expertise areas" in which new product development or other innovations may be expected. Basic research projects are normally allowed a period of 5 years in which to complete a certain task. Although no formal review procedures are applied during this period, the "nuggets" of basic research are presented to the companies for comment once a year.

The elaborate programming of Multifulcrum's R & D illustrates the considerable amount of time and effort such a company invests in ensuring that R & D is user-controlled. The strong participation of companies in the formulation of research projects and the fact that the companies pay for the projects ensure that technology development is consumer-oriented. Elaborate linkage mechanisms (committees, special officers, "body swapping", etc) ensure frequent interaction, involvement of all interested parties and careful decision making.

Placing such emphasis on user control is essential for profit-making organizations, whose profits are a direct reflection of consumer reaction to the products of their R & D departments. Publicly funded organizations such as government or university research institutions, whose feedback about their performance from their clients is not spelt out in financial terms, tend to be less client-oriented. But they can ill afford to be. In short, user control would appear to be a key organizational principle for an effective AKIS.

User control of major processes within the AKIS is an essential attribute if default conditions are to be avoided.

In many developing countries, it might be more cost-effective to create user control by giving farmers control over part of the AKIS budget, than to strengthen research/extension services. To develop the AKIS by investing in laboratories, staff training, cars, megaphones and management capacity for its research/extension services is to behave like the marriage counseller who tries to save a marriage by giving the strongest partner assertiveness training.

The Limitations of User Control

Having made such a strong plea for user control, I should in fairness discuss some of its drawbacks.

Strong farming lobbies can pull agricultural policy in directions which are detrimental to national development. Worse, the fact that the farmers who exert control are often resource-rich and atypical can lead to policies and production systems which are exploitative and a serious barrier to the development of small-scale farming. User control in itself is no guarantee of equity. If the AKIS is to serve categories of farmers other than the resource-rich, strong pressure must be exerted by management, policy groups, donors or other prime movers to give such categories user control also.

Moreover, if users' economic interests conflict with government policy in areas such as the environment, increased user control will only exacerbate existing problems.

Bridging the Gap

The medium and small business sector provides an interesting comparison with the resource-poor farming sector because it too is not homogeneous enough to share a single knowledge and information system. It is too fragmented to be able to benefit from a joint R & D effort, as do the production companies in a large multinational corporation such as Multifulcrum. Elaborate external support programs — analogous to research/extension services in the agricultural sector — have been developed to perform R & D functions for such businesses but, as in agriculture, these often do not make much use of such services.

Process consultation: A useful tool. An alternative approach is to help such businesses become more innovative, that is, to enhance their capacity to make use of external information, instead of simply pushing more innovations at them (for example, Buijs, 1984 and 1987). This was the aim of an experimental project launched in the Netherlands. The main strategy of the Industrial Innovation Project was not to find innovations for the participating firms, but to teach them how to become more innovative. The project's consultants were trained in what they called "process consultation", which involved introducing a step-by-step model of the innovation process, stimulating the creativity of company staff members and encouraging the use of external information (Buijs, 1984 and 1987).

In agriculture, as in industry, process consultation is a useful concept complementing the existing role played by expert consultation. While the latter provides an external input of information or technology, the former is a means of mobilizing people — educating them and organizing them to become more effective participants in the AKIS. The complementary nature of the two aspects becomes obvious in such contexts as the history of agricultural development in Israel (Blum, 1987), where the need to develop agriculture from scratch with often totally inexperienced farmers, or to change course from producing national food crops to producing specialized export crops, amply illustrated the importance of combining the external input role with an internal educational and organizational one. However, despite the need for both processes, most AKIS in the developing world remain weak in process consultation.

To sum up, the input of information from external sources is crucial to the effective functioning of any knowledge and information system, but if the system does not have the capacity to generate and enhance appropriate roles for its constitutent parts, it will not be in a position to absorb such information. In the agricultural sectors of developing countries, the lack of social organization among small-scale farmers is therefore a considerable barrier to development.

Improving linkage mechanisms. A linkage mechanism is the concrete procedure, regular event, arrangement, device or channel which bridges the gap between components of a system and allows communication between them. The term linkage mechanism should be distinguished from interface. Although in computer terminology, "interface" is used for a "device linking two systems" (Hurthubise, 1984), it is preferable to reserve "interface" for the "force field" between two institutions. The linkage mechanism is the device which operationalizes the interface (Engel, pers. comm.).

In some countries, the annual report of the research institute — often published late and sometimes not published at all — is still virtually the only official linkage mechanism between research and technology transfer. In others, much has been done in recent years to increase the number of linkage mechanisms and improve their effectiveness. Generally speaking, the greater the number of linkage mechanisms, and the greater the range they span within the administrative hierarchy, the better the chances that effective links will develop.

A typical example of a project with multiple linkage mechanisms is the Ghana Grains Development Project, which has:

1. *A survey carried out jointly by research and extension staff:* at the beginning of every cropping season, a team of breeders, economists, agronomists and extension staff pay informal visits to farms and ask questions about farmers' problems. The answers are used to draft a questionnaire, and on the basis of the results of the questionnaire, research projects are formulated for that year.
2. *Quarterly meetings of the members of the on-farm, economic and extension programs:* discussion at these meetings centers on current trials and surveys, and on the plans for the following quarter.
3. *Annual reports:* published regularly, these describe the various research programs and their results. They are intended mainly for a scientific audience.
4. *An annual workshop:* at this workshop all the year's research and extension activities are presented to a large audience. It is attended by members of the AKIS from all parts of the country and by representatives from foreign research institutes.
5. *A pre-workshop meeting:* before the annual workshop all senior officers of the project meet in order to transform research findings into recommendations. A committee is then charged with the responsibility of turning the recommendations into comprehensible language for use by extension officers and literate farmers. The result is a booklet entitled *Maize and Cowpea Guide*. The booklet is updated before every workshop, to keep abreast of current findings.

6. *Training programs:* research officers of the project explain in detail the latest recommendations on the crop to field agents organized in groups according to agro-climatological zones. Issues encountered in the field are raised by extension workers.
7. *Field days:* these are organized three times during a planting season. Research officers normally participate, thus gaining first-hand experience from both the farmers and the field extension workers (Annor-Frempong, 1988).

Meetings, written communications and joint activities are all examples of formal linkage mechanisms, but informal mechanisms such as sharing a coffee room, or playing golf together, may also be just as important.

As we have already seen, an effective formal mechanism used in industrial corporations is "body swapping". After having worked in one department, say basic research, a researcher may be posted to the R & D department of a company. Body swapping creates communication bridges, allowing informal contact between different subsystems. An analogous mechanism used in the agricultural sector of some developing countries is the post of Research-Extension Liaison Officer. These officers are recruited from extension services to work in on-farm adaptive research teams; they play an especially important role in enlisting the support of extension services once technology is ready for more widespread testing and dissemination.

Although linkage mechanisms have become both more varied and more sophisticated in recent years, there are some interfaces which no linkage mechanism can bridge. The status or cultural differences between two institutions may be too great, their goals may be too different, their competition for the same resources may be too keen, or the span in the calibration of the research-practice continuum may be too long (not enough cogs in the gearbox, *see* Calibrating the science-practice continuum). These problems should be emphasized, because it is often believed that communication problems can be solved by imposing some new linkage mechanism on the two institutions. Experience shows that such artificially introduced linkage mechanisms do not work unless the interface allows it.

Optimum heterophily. The twin concepts of homophily and heterophily are a useful tool for looking at some of these problems. According to Rogers (1983), there must be some similarity in culture, language, socio-economic status, etc, before communication between two institutes can take place. People tend to communicate most frequently and effectively with those who are most similar to themselves. Rogers coined the term "heterophily gap" to describe a difference between two parties attempting to communicate which makes it difficult for them to do so. On the other hand, people whose jobs are too similar may have little information to exchange. The optimum conditions for good communication would appear to be loosely related fields of work shared by people of similar outlook. These conditions are known as "optimum heterophily". Experience shows that heterophily between graduates and non-graduates is very difficult to overcome.

Often, the only communication between research and extension is that between researchers and senior extension officers. Subject-matter specialists, though explicitly appointed in a liaison function, may not be able to play that role if the status gap is too large (Blok and Seegers, 1988). Linking a graduate in the research department with a graduate in the extension department is, of course, no guarantee that the link will be effective: within the extension department, the same gap between graduates and non-graduates may exist.

Calibrating the science-practice continuum. The processes or functions of an AKIS can be said to be distributed along a science-practice continuum (Lionberger and Chang, 1970). If information and technology are to flow smoothly from one part of the AKIS to another, this continuum must be finely calibrated. The graduations in the sequence "research-extension-users" are too coarse: they do not allow for the "scaling down" required to bridge the gaps between institutions. This scaling down can be compared to the transmission of a car. To be drivable, a car needs several gears representing different speeds, each of which must be used in sequence. The gear cogs are designed to allow smooth transmission from one speed to the next. Linkage mechanisms (synchronization, double declutching) are still necessary to bridge the speed gaps between the cogs.

Many authors have defined the required sequence of cogs on the science-practice continuum in terms of the "functions" (or "steps" or "stages") which must be performed (for example, Beal and Meehan, 1986; Havelock, 1986; Lionberger, 1986; Rothman, 1986). For the most part, they have described a downstream sequence using terms such as basic/strategic research, technology generation, technology testing, technology adaptation, technology integration, dissemination, diffusion and adoption (McDermott, 1987; *see* Figure 4 *overleaf*). McDermott clearly distinguishes between these "functions" and the existing institutions of research and extension. He claims that the responsibilities conventionally assigned to existing research and extension organizations may leave a "fatal gap" in the performance of these functions:

> In even the best of cases, research often stops midway through the testing process. Testing is not finished until it is done in the systems in which the technology is expected to perform. At the other end of the continuum, extension does not expect to start until the dissemination function. The seriousness of the gap is apparent.

Figure 4. Functions of the AKIS

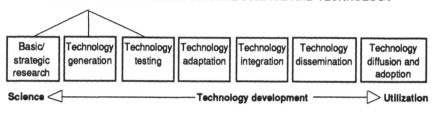

Source: Adapted from McDermott (1987)

This gap is often encountered in practice. A typical example is provided by the QDPI. This highly sophisticated AKIS recognizes three basic policy instruments: research, extension and regulation. The research divisions focus on publications as their main product. It is for publishing that researchers are rewarded. However, publishing does little to promote technology transfer. Extension is supposed to provide services to farmers and disseminate innovations. The gap is obvious and generally recognized by Queensland extension staff: there is no "development function". A number of extension staff have started to do development work themselves, but there are no budget and facilities for it, and no reward is offered for doing it.

McDermott continues:

Farming systems research is providing an exceptionally effective means by which research can move into that gap from its end of the process and effect the interaction with farmers. As of now, extension has not made a significant move into the gap from its end of the process.

Various types of institutional go-between, such as subject-matter specialists, or technical liaison officers and supporting staff, are currently developing to bridge the gap. The job descriptions of these go-betweens are still evolving—often there is difficulty in defining them—and it is still too early to say whether or not they will be successful. Their main functions are: (1) to maintain liaison with research so as to keep abreast of new technical developments and help translate field problems into researchable questions; (2) to establish links with input suppliers so as to improve the chances that necessary inputs will be available; and (3) to provide technical support to field staff and pick up field problems from them. Important functions of technical liaison are adaptive research, training, developing reference materials and training aids, trouble shooting, and responding to extension agents' requests for help.

Clearly, the fatal gap operates in both directions on the continuum. However, it is probably even larger in the upstream than in the downstream direction. Returning to Figure 4, we can see that we at least have names for the functions required in the downstream flow from research to users, even if not all the functions are performed. The functions required to ensure the upstream flow of needs and problems have not even been described. We have already seen how few of the problems expressed by producers can be turned into problems that can be addressed by research. Clearly, we have little insight as yet into the nature of bottom-up flows of communication at each transformation point on the science-practice continuum, nor of the interaction between top-down and bottom-up flows. According to Biggs, it is often difficult to get even minimal funds and time allocated to the process of learning from farmers who might positively influence the priorities and programs of experiment stations (Biggs, 1983 and 1989).

One of the major problems in achieving an improved flow of information in both the upstream and the downstream directions is the dual allegiance of the extension service in many developing countries. As Wagemans (1987) discovered, the strong client orientation of village-level extension workers often conflicts with the policy orientation of their bosses, motivating the former to avoid providing the latter with accurate information. At the interface between the two, rituals occur which are designed to ensure that bosses continue to believe their instructions are being carried out. As a result, the extension service may develop a "split personality", in which one half scarcely even recognizes the existence of the other. Under these conditions it becomes impossible both to assess the output of the extension service and to implement policy directives through it.

Domain correspondence. One of the major problems in bridging the gap between research and technology transfer is that the categories according to which work is organized often differ. These categories, or "domains" as they are called in the literature, may be research disciplines (for example, soil science), sectors (for example, dairy farming), commodities (for example, cocoa), agro-ecological zones (for example, derived savannah zone), geopolitical areas (for example, provinces or districts), industries (for example, the beef cattle industry), farm classes (for example, middle-sized farms), and so on. Usually, the domain used changes as one moves along the continuum from science to practice. Thus strategic and applied researchers often form commodity teams, transforming disciplinary knowledge into crop-specific knowledge, while technology transfer workers are frequently organized according to industry or geopolitical area. It is difficult to make the different domains relate to each other.

The QDPI is an interesting case because it has divided up the whole AKIS into industry- or service-oriented branches, irrespective of particular activities, disciplines, farming systems, environments or other categories. Examples of these branches are the Beef Cattle Production Branch, the Horticulture Branch, the Veterinary Services Branch and the Crop Production Branch. The Director of each branch holds a powerful headquarters position and ranks just below the Director General. The branches have their own research, extension and regulatory personnel, extending right down to field level.

Communication across branches is not always easy. In the regions, staff of different branches are often housed in the same building, and some comradery and cooperation usually develop at that level. However, the real problem occurs at the farm level. A beef cattle producer needs integrated information from the Beef Cattle Production Branch, the Veterinary Services Branch, the Pasture Development Branch and the Economic Services Branch — integrated, that is, from the point of view of his cattle enterprise as a profit-making business. The solution to this problem has been to appoint cross-branch Regional Extension Leaders, reporting directly to the Deputy Director General (a form of matrix management). Their "nudging" role has been to create "industry groups" consisting of representatives of different branches at the farm level (Röling et al., 1987).

AKIS Disorders

From an analysis of existing AKIS, a number of major AKIS disorders can be identified. Here are 12 of them; all sources appear in this paper and/or in the reference list, apart from van Dissel (pers. comm.) and Mansholt (former Dutch Minister of Agriculture).

1. Engel's *wrong plugs:* the lack-of-fit between the domains used by different components (for example, a commodity research institute is used to backstop extension servicing a multi-crop farming system)
2. McDermott's *fatal gap:* a functional gap which cannot be bridged by linkage mechanisms because of insufficient calibration of the science-practice continuum (for example, the non-existence of adaptive research because applied research is satisfied with producing scientific publications, while extension starts from the recommendation)
3. Biggs' *mis-anticipation:* the lack-of-fit between the conditions anticipated during technology development and those in which technology is used (for example, the formulation of fertilizer recommenda-

tions on the basis of crop responsiveness under unrepresentative research station conditions)

4. van den Ban's *cross-purpose disability:* reward systems and incentives which encourage AKIS components to reduce their synergy (for example, rewarding members of the Research Division for scientific publication and not for producing innovations for farmers)

5. Wagemans' *flow blocks:* lack of effective linkage mechanisms through which information can flow (for example, a "split personality" in the extension hierarchy, whereby lower levels become strongly service-oriented while higher levels are subject to policy directives, preventing effective communication between them)

6. Mosher's *mix insufficiency:* the lack of provision of one or more of the conditions which are essential for technology utilization (for example, research-based recommendations for farmer behavior, while the required inputs are not available)

7. Ascroft's *equity syndrome:* progressive farmer control of the AKIS, biasing technology development in favor of a minority of farmers

8. van Dissel's *policy folly:* the use of knowledge and information as if they were policy instruments with compulsive power, especially if coupled with disregard for the fact that farmers have to live by their results and extension workers are in daily contact with farmers

9. Rogers' *heterophily gaps:* interfaces between components which differ so greatly that linkage mechanisms cannot span them

10. *No juice:* blaming the AKIS when the problem is agricultural prices

11. Mansholt's *small farm squeeze:* setting in motion technology-driven agricultural development in high-potential areas and for resource-rich farmers without regard for the employment and livelihood effects on small-scale farmers and those in less well endowed areas

12. Jiggins' *out-a-synch:* giving priority to knowledge system development when the majority of farmers have not gained sufficient control over their production environment to use a regular flow of new knowledge

Diagnosing these disorders in a given AKIS is only a first step toward curing them, but it is an important one.

Conclusions

Managers participating in the management of AKIS in developing countries face a very difficult challenge. I hope the conclusions presented here, based largely on Röling and Engel (1988), will be helpful to them.

If agricultural development is to be enhanced, there is a need to nudge widely differing institutions, often under different administrative arrangements, both public and private, into compatible roles. This task requires the introduction of some shared model of the local AKIS, as well as an understanding of each institution's role in the system (van Beek, 1988). A shared model is a prerequisite for effective links between research and technology transfer.

The effective management of information requires the design, differentiation and development of the range of institutions involved in such functions as generation, exchange, transformation, integration, dissemination and utilization of knowledge and information, in such a way that the AKIS operates synergically. Technology development is a crucial task of the system.

Earlier models of the AKIS (Nagel, 1980), in which specific functions (for example, knowledge generation) were assigned to specific institutions (for example, research), must now make way for new ones which allow for the fact that all major parties in the system engage in all its major functions (Engel, 1987).

Also required is a detailed understanding of the functions which are to be performed by the system. The AKIS manager must have a thorough grasp of these functions in order to assess the adequacy of the present institutional structure and orientation. Usually, strategic, applied and adaptive research are identified as appropriate functions of the AKIS, as well as technology integration, transfer and use. However, the functions identified depend upon the underlying assumptions which are made about the type of model to be used. There is still too much emphasis on "downstream" functions, suggesting the use of a one-way model. Other than "feedback", which denotes reaction rather than action, we have no words for the functions to be performed in shifting indigenous knowledge and farmer influence "upstream" towards the science end of the science-practice continuum.

The interfaces between each of the system's major entities are especially vulnerable. Most of the disorders to which AKIS are subject, such as problems caused by conflicting domains, heterophily gaps, ineffective linkage mechanisms and so forth, occur at the interfaces. Faulty interfaces lead to failure to transform knowledge and information appropriately, and hence to a system that cannot operate synergically. The management of AKIS interfaces is therefore a crucial task — indeed, some observers see it as the most essential task of all.

A promising tool for managing interfaces is a matrix of all the relevant research, technology transfer and user entities in the local AKIS. Each cell of the matrix represents an interface. These interfaces can be weighted according to management criteria such as frequency of use, or importance to the system as a whole. In this way, rational decisions regarding the allocation of time, attention and financial resources to each interface can be made (van Beek, 1988).

Another important task is to ensure a balance between the power of the various institutions to intervene, and the countervailing power of the various categories of users to control the character of those interventions. A hallmark of successful AKIS is that users have considerable control over technology development and transfer. Such control is not yet strong enough in many AKIS in developing countries, and this is one of the major factors leading to systems that operate under default conditions. Most AKIS in developing countries would probably benefit more by increasing the countervailing power of their clients than by increasing the power to intervene of research and extension.

Technological innovation is considered essential for achieving food security or self-sufficiency, competitiveness in foreign markets, efficient use of resources and other national goals. However, technology-driven development usually marginalizes small-scale farmers, destroys ecosystems and incurs foreign exchange costs (imports of capital goods and chemicals) as well as bringing benefits.

A crucial task is therefore to ensure that the intended categories of clients are in fact being served by the AKIS. In most AKIS, a minority of progressive farmers has undue influence and is able to ensure that the system meets their needs, thereby often by-passing the needs of resource-poor farmers. Avoiding this default condition requires the deliberate targeting of technology development toward the latter group, who must simultaneously be given the ability and power with which to articulate and press their needs.

Privatization is a partial solution to the problem of funding agricultural research and development, and one that is increasingly being used in such widely differing countries as the Netherlands and China (Delman, 1988). However, only the service parts of the AKIS are suitable for privatization: the problem of finding public sector support for research remains, because private sector R & D is inevitably targeted toward a paying clientele — the more progressive farmers. In addition, privatization complicates the task of linking research and technology transfer.

The nature of the resource base has special implications for the management of the AKIS. In large, simple, homogeneous farming systems producing a large surplus for urban consumption or export (for example, the Dutch dairy industry, irrigated rice in Thailand, wheat in northern France), expenditure on the AKIS has a high pay-off per dollar invested. The scale of the production system allows a large dedicated R & D establishment, focusing on a single commodity or industry. It is easy, under these conditions, to achieve integration between input delivery, primary production, processing, marketing and so on.

However, in complex, diverse, small-scale farming systems that are not producing a large surplus, a dedicated and effective AKIS cannot easily evolve. This is the case for most rainfed farming systems in developing countries, where farmers' ability to control the production environment is usually insufficient. What is first required is investment in a permanent, stable production capacity through such technologies as terracing, water harvesting and afforestation (Haverkort, 1988). Rainfed farming systems must be knowledge-intensive if those who obtain their living from such systems are to deal with the increasing complexity that surrounds their management. So far, the challenge of developing a suitable AKIS for rainfed farming systems has not been met satisfactorily.

Although an AKIS may consist of many autonomous or semi-autonomous organizational units, it needs to be managed as if it were a seamless whole. Only then will it perform synergically.

Acknowledgments

I would like to thank Dr David Kaimowitz and Paul Engel for their assistance in drafting this paper, and David Carrigan, Mervin Littman and Peter van Beek of the QDPI, and Dr Shankariah Chamala of Queensland University, for helping Dr Janice Jiggins and me, during our consultancy at QDPI in 1987, understand the research and extension structure in Queensland. Special thanks are due to Drs S. Bruin, S. Gorter and J. Geurts, all of whom work for multinational corporations, for acquainting me with fascinating aspects of R & D management in large corporations.

I also wish to thank Mahinda Wijeratne, Charles Annor-Frempong, Kees Blok, Stephan Seegers, Mangala de Soyza, Anje de Ruiter, Miriam Langen and Carine Coehoorn, who carried out empirical research as Ph.D or M.Sc students at the Wageningen Agricultural University and whose work I have been able to draw on. I am very grateful for the comments of my colleagues Dr Janice Jiggins, Dr Anne van den Ban, Dr Abraham Blum, Dr Herbert Lionberger and Dr Loek de la Rive Box. Dr Paul Macotte and Dr Ajibola Taylor of ISNAR, and Simon Chater of Chayce Publication Services, made important contributions to the structure and readability of the paper. Any remaining shortcomings are my own responsibility, however.

References

Annor-Frempong, C. "An evaluation of the research/extension interface: The case of cocoa and maize in Ghana." M.Sc thesis. Wageningen: Agricultural University of Wageningen, 1988.

Arce, A., and Long, N. "The dynamics of knowledge interfaces between Mexican agricultural bureaucrats and peasants: A case study from Jalisco." *Boletin de Estudios Latinamericanos y el Caribe* (December, 1987).

Ascroft, J. "Modernization and communication: Controlling environmental change." Ph.D thesis. East Lansing: Michigan State University, 1972.

Bantje, H. "Household differentiation and production: A study of small-holder agriculture in Mbozi District." Dar-es-Salaam: University of Dar-es-Salaam, 1987.

Beal, G.M., Dissanayake, W., and Konoshima, S. (eds). *Knowledge Generation, Exchange and Utilization.* Boulder: Westview Press, 1986.

Beal, G.M., and Meehan, P. "Communication in knowledge production, dissemination and utilisation." *Knowledge Generation, Exchange and Utilization.* Beal, G.M., Dissanayake, W., and Konoshima, S. (eds). Boulder: Westview Press, 1986.

Bennett, C. "Improving coordination of extension and research through the use of interdependency models." Paper presented at the Seventh World Congress of the International Rural Sociology Association, Bologna, 1988.

Berlo, D. *The Process of Communication.* New York: Holt, Rinehart and Winston, 1960.

Biggs, S. "Monitoring and control in agricultural research systems: Maize in northern India." *Research Policy* (12, 1983).

Biggs, S. "Resource-poor farmer participation in research: A synthesis of experiences from nine national agricultural research systems." OFCOR Comparative Study No 3. The Hague: ISNAR, 1989.

Blok, K., and Seegers, S. "The research-extension linkage in the southern region of Sri Lanka: An agricultural information systems perspective." M.Sc thesis. Wageningen: Agricultural University of Wageningen, 1988.

Blum, A. "The Israeli experience in agricultural extension and its application to developing countries." *Agricultural Extension Worldwide.* Rivera, B., and Schram, S. (eds). New York: Croom Helm, 1987.

Bos, J., and Burgers, W. "Voorlichters en veranderingen in hun werk." M.Sc thesis. Wageningen: Agricultural University of Wageningen, 1982.

Bronsema, H.J., and van Nimwegen, N. "Menselijke mierenhopen en de verstedelijking van de wereld." *Demos* (3, 1987).

Buijs, J.A. "Innovation can be taught." *Research Policy* (vol 16, no 6, 1987).

Buijs, J.A. *Innovatie en Interventie.* Deventer: Kluwer, 1984.

Bunting, A.H. "Extension and technical change in agriculture." *Investing in Rural Extension: Strategies and Goals.* Jones, G.E. (ed). London: Elsevier, 1986.

Castillo, G.T. "Technology transfer in an imperfect world." Lecture Notes. Wageningen: Agricultural University of Wageningen, 1983.

Chambers, R. *Rural Development: Putting the Last First.* London: Longman, 1983.

Chambers, R., and Jiggins, J.L.S. "Agricultural research for resource-poor farmers: A parsimonious paradigm." *Agricultural Administration and Extension* (27, 1987).

Collinson, M. "Farming systems research: Diagnosing the problems." *Research, Extension, Farmer: A Two-Way Continuum for Agricultural Development.* Cernea, M.M., Coulter, J.K., and Russell, J.F.A. (eds). Washington D.C.: World Bank, 1985.

Cordova, V., Herdt, R.W., Gascon, F.B., and Yambao, L. "Changes in rice production technology and their impact on rice farm earnings in Central Luzon, the Philippines, 1966-79." Departmental Paper. Los Banos: IRRI, 1981.

Delman, J. "The agricultural extension system in China." Unpublished paper. Aarhus: University of Aarhus, 1988.

de Vries, A. "Training farmers in agricultural research." Ph.D thesis (in preparation). Wageningen: Agricultural University of Wageningen.

de Soyza, M. "Farmer-instigated agricultural research in Sri Lanka." M.Sc thesis. Wageningen: Agricultural University of Wageningen, 1988.

Elliott, H. "Diagnosing constraints in agricultural technology management systems." Paper presented at the International Workshop on Agricultural Research Management, ISNAR, The Hague, 1987.

Engel, P. "Rural development and the transformation of agricultural knowledge." Plenary Lecture for International Course on Rural Extension, International Agricultural Center, Wageningen, 1987.

Foster, G.M. *Traditional Cultures and the Impact of Technological Change.* New York: Harper and Row, 1976.

Fresco, L.O. *Cassava and Shifting Cultivation: A Systems Approach to Agricultural Technology Development.* Amsterdam: Royal Tropical Institute, 1986.

Havelock, R.G. "Planning for innovation through the dissemination and utilisation of knowledge." Mimeo. Ann Arbor: University of Michigan, 1969.

Havelock, R.G. "Linkage: A key to understanding the knowledge system." *Knowledge, Generation, Exchange and Utilization.* Beal, G.M., Dissanayake, W., and Konoshima, S. (eds). Boulder: Westview Press, 1986.

Haverkort, A.W., and Röling, N.G. "Six rural extension approaches." Paper presented at the International Seminar on Strategies for Rural Extension, International Agricultural Center, Wageningen, 1984.

Haverkort, A.W. "Agricultural production potentials: Inherent, or the result of investments in technology development?" *Agricultural Administration and Extension* (30, 1988).

Hubbert, C.A. "Postharvest extension on mangoes." Mimeo. Hamilton: QDPI, 1987.

Hurthubise, R. *Managing Information Systems: Concepts and Tools.* West Hartford: Kumarian Press, 1984.

International Rice Research Institute. "Integrated measurement of the environmental and socio-economic impact of new technology in rice farming areas." Paper presented at the Expert Consultation for Identifying Issues and Developing Methodologies for Measuring the Environmental, Socio-economic, and Interactive Impacts of New Rice Technology, Los Banos, 1987.

Jiggins, J. "Gender-related impacts and the work of the international agricultural research centers." CGIAR Study Paper No 17. Washington D.C.: World Bank, 1986.

Jiggins, J. "Conceptual overview: How poor women earn income in rural sub-Saharan Africa and what prevents them from doing so." Nairobi: Ford Foundation, 1988.

Kaimowitz, D., Snyder, M., and Engel, P. "A conceptual framework for analyzing the links between research and technology transfer in developing countries." Linkages Theme Paper No 1. The Hague: ISNAR, 1989.

Lionberger, H.F., and Chang, H.C. *Farm Information for Modernizing Agriculture: The Taiwan System.* New York: Praeger Press, 1970.

Lionberger, H.F. "Towards an idealised systems model for generating and utilizing information in modernizing societies." *Knowledge Generation, Exchange and*

Utilization. Beal, G.M., Dissanayake, W., and Konoshima, S. (eds). Boulder: Westview Press, 1986.

Lipton, N., and Longhurst, R. "Modern varieties, international agricultural research and the poor." CGIAR Study Paper. Washington D.C: World Bank, 1985.

Martínez Nogueira, R. "The effect of changes in state policy and organization on agricultural research and extension links: A Latin American perspective." Linkages Theme Paper No 5. The Hague: ISNAR, 1989.

McDermott, J.K. "Making extension effective: The role of extension/research linkages." *Agricultural Extension Worldwide.*. Rivera, W., and Schram, S. (eds). New York: Croom Helm, 1987.

Mosher, A.T. *Getting Agriculture Moving.* New York: Agricultural Development Council, 1966.

Nagel, U.J. "Institutionalization of knowledge flows." *Quarterly Journal of International Agriculture* (30, 1980).

Rogers, E.M., Eveland, J.D., and Bean, A.S. "Extending the agricultural extension model." Mimeo. Stanford: Stanford University Press, 1976.

Rogers, E.M., and Kincaid, L.D. *Communication Networks: Towards a New Paradigm for Research.* New York: Free Press, 1981.

Rogers, E.M. *Diffusion of Innovations.* New York: Free Press, 1983.

Rogers, E.M. "Models of knowledge transfer: Critical perspectives." *Knowledge Generation, Exchange and Utilization.* Beal, G.M., Dissanayake, W., and Konoshima, S. (eds). Boulder: Westview Press, 1986.

Röling, N. "The evolution of civilization." Ph.D thesis. East Lansing: Michigan State University, 1970.

Röling, N. "Extension science: Increasingly preoccupied with knowledge systems." *Sociologia Ruralis* (25, 1986a).

Röling, N. "Extension and the development of human resources: The other tradition in extension education." *Investing in Rural Extension: Strategies and Goals.* Jones, G.E. (ed). London: Elsevier, 1986b.

Röling, N. "Knowledge utilization: An attempt to relativate some reified realities." *Knowledge Generation, Exchange and Utilization.* Beal, G.M., Dissanayake, W., and Konoshima, S. (eds). Boulder: Westview Press, 1986c.

Röling, N. *Extension Science: Information Systems in Agricultural Development.* Cambridge: Cambridge University Press, 1988a.

Röling, N. "Agricultural knowledge systems: The context for research/extension linkage." Paper presented at the Seventh World Congress of the International Rural Sociology Association, Bologna, 1988b.

Röling, N., and Engel, P. "IKS and knowledge management: Utilizing indigenous knowledge in institutional knowledge systems." Paper presented at Conference on Indigenous Knowledge Systems: Implications for Agricultural and International Development, Academy for Educational Development, Washington, 1988.

Röling, N., Jiggins, J., and Carrigan, D. "Extension as part of an agricultural knowledge system." Paper presented at the Australasian Agricultural Extension Conference, Brisbane, 1987.

Rothman, J. "The research and development model of knowledge utilization: Process and structure." *Knowledge Generation, Exchange and Utilization.* Beal, G.M., Dissanayake, W., and Konoshima, S. (eds). Boulder: Westview Press, 1986.

Safilios Rothchild, C., and Mburugu, E.K. "Men's and women's agricultural production and incomes in rural Kenya." Paper presented at the Seminar on Agricultural Development, Population and the Status of Women, Nyeri, Kenya, 1987.

Sands, C.M. "The theoretical and empirical basis for analyzing agricultural technology systems." Ph.D thesis. Champaign-Urbana: University of Illinois, 1988.

Sheridan, M. "Peasant innovation and diffusion of agricultural technology in China." Rural Development Study. Ithaca: Cornell University Press, 1980.

Sims, H., and Leonard, D. "The political economy of the development and transfer of agricultural technologies." Linkages Theme Paper No 4. The Hague: ISNAR, 1989.

Sofranko, A., Morgan, G., and Kahn, A. "Insights into farmer-extension contacts: Evidence from Pakistan." Paper presented at the Seventh World Congress of the International Rural Sociology Association, Bologna, 1988.

Steinberg, D.I., Jackson, R.I., Kim, K.S., and Hae-kyun, S. "Korean agricultural research: The integration of research and extension." Project Impact Evaluation. Washington D.C.: USAID, 1982.

Swanson, B. "INTERPAKS and the development of its knowledge system model." Plenary Lecture for the International Course on Rural Extension, International Agricultural Center, Wageningen, 1986.

Swanson, B. and Claar, J. "Technology development, transfer and feedback systems in agriculture: An operational systems analysis." Unpublished paper. Champaign-Urbana: University of Illinois, 1983.

van Beek, P. "The Queensland dairy AKIS: An application of the systems approach to the management of research and extension units in the dairy industry in Queensland, Australia." M.Sc thesis. St Lucia: University of Queensland, 1988.

van den Ban, A.W., and Hawkins, S. *Agricultural Extension*. Harlow: Longman, 1988.

van der Ploeg, J.D. "De verwetenschappelijking van de landbouwbeoefening." Wageningen: Agricultural University of Wageningen, 1987.

Wagemans, M. *Voor de Verandering*. Ph.D thesis. Wageningen: Agricultural University of Wageningen, 1987.

Wijeratne, M. "Extension for rice farmers in Sri Lanka: An empirical study of the knowledge systems with special reference to Matara District." Ph.D thesis. Wageningen: Agricultural University of Wageningen, 1988.

World Bank. *World Development Report*. Washington D.C.: World Bank, 1982.

World Commission on Environment and Development. *Our Common Future*. Oxford: Oxford University Press, 1987.

Wuyts, A. "Agricultural research-technology transfer system: The People's Republic of China." ISNAR Staff Note. The Hague: ISNAR, 1988.

2

The Political Economy
of the Development and Transfer
# of Agricultural Technologies

Holly Sims and David Leonard

This paper considers the effectiveness of institutional agricultural technology systems (IATS) in relation to four criteria: integration, relevance, responsiveness and adoption. Integration refers to the extent to which agricultural research is integrated with technology transfer. Relevance refers to the capacity of an IATS to develop technologies relevant to farmers' most pressing needs. Responsiveness denotes an IATS' ability to respond to the poor majority of farmers, who lack financial resources and favorable agro-ecological conditions. Adoption is a derivative variable which reflects the usefulness of technologies, success in meeting farmers' needs, and the "marketing" capability of the system. The four criteria are clearly interrelated, but performance in relation to each of them varies in strength within each IATS. Thus it is useful to keep all four criteria in mind as we proceed.

We will begin by discussing the external factors which inhibit integration, relevance and responsiveness, and which deflect attention from adoption. Our focus will be on two major sets of factors: those arising from historical institutional legacies, and those related to the contemporary socio-political structures of most low-income countries.

In the absence of external pressures to meet the four criteria, institutions generally follow internal dynamics, with the result that performance is poor. We continue by developing a stylized "model" of what research and extension look like under these default conditions.

Four groups of people can exert forces which might lead to better integrated and more effective IATS: national policy makers, external donor agencies, farmers' organizations and commercial firms. We will end by assessing the patterns of improved performance these groups can elicit.

Inhibiting Factors

Historical Institutional Legacies

Most public sector IATS in developing countries today owe their origins to colonial powers and foreign commercial interests. These actors left an imprint on institutions and values which militates against integration, relevance and responsiveness. To illustrate, we will concentrate on Asian and African countries colonized by the British. In many respects, their experiences were shared by societies which were not colonized, or were colonized by other European states.

Colonial powers established research institutes to increase the production of high-value export commodities. Answerable to government departments in Europe, these institutes did not serve the majority of subsistence producers. To encourage the production of export crops, policy makers extended incentives to the small proportion of the rural population that could afford to divert land away from the production of subsistence crops. That proportion typically included foreign settlers and companies, and indigenous landed elites. Such producers were eager for innovations which would increase their profits, and they established a close relationship with researchers.

Scientists were responsive to the needs of these elite farmers and developed knowledge relevant to their conditions. Formal channels to popularize innovations were largely unnecessary. Interchange was facilitated by the socio-economic status of producers and by their limited numbers, which allowed them to communicate easily and directly with researchers. It was professionally rewarding for researchers to work for clients whose farms were comparable to research stations, for scientists could minimize the number of variables to be considered and solve problems quickly.

In many cases, export crop producers included a relatively advantaged segment of peasant farmers. Yet the relationships of professionals with these peasant farmers tended to be top-down and coercive, in contrast to their relationships with elite producers. Researchers did not address the infinitely more complex problems of the vast majority of resource-poor subsistence farmers (Chambers and Jiggins, 1987).

The need for mechanisms to develop scientific findings for a broader clientele and disseminate them more widely was eventually recognized by British colonial policy makers, but considerably less so by their French counterparts. British administrators had a keener appreciation of government's responsibility to put bread on the table, perhaps because Britain's domestic food supply was more precarious than that of France. In addition, the threat of famine in some of Britain's populous Asian colonies compelled policy attention to the problem of subsistence food production.

British colonial officers generally charged a Department of Agriculture with responsibility for extension activities to translate research findings into simple language for a mass audience. (There was a strong element of coercion associated with these extension activities, exemplified by some soil conservation efforts and resettlement programs). Separate bureaucratic agencies were established for crops and animal husbandry. They were not integrated with one another, nor were they closely associated with research bodies (von Blanckenburg, 1984). The French generally established semi-autonomous parastatal agencies responsible for particular crops. Such bodies were more integrated than the British system, but even less concerned with the needs of subsistence food producers.

Integration of research and technology transfer for subsistence crops was impeded by marked differences in status between professional research staff and technology transfer agents, and between the latter and producers. Those with subordinate status were seen as passive recipients of information. Since scientists had limited knowledge of subsistence agriculture, research was seldom relevant to the rural majority. Farmers relied more on informal research and technology transfer between each other. Extension staff were expected to promote modern European farming practices, not to interpret the farmers' world to researchers.

Extension agents were often held accountable for assignments not necessarily related to agriculture. As the sole representatives of officialdom at the local level, they performed a variety of functions including data collection, mobilization of political support and maintenance of order. Their residual advisory work was not grounded in solid research geared to clients' circumstances, and so it was not valued by farmers. It is not surprising that technology development and transfer yielded scant returns in terms of agricultural productivity.

Many new states emerged from colonialism with a limited number of research institutions concerned with export commodities, but few or none concerned with food crops or other commodities produced by resource-poor farmers. Where the latter existed, they were loosely coupled with technology transfer mechanisms through one-way communication chains.

This pattern has been reinforced in the postcolonial era by a number of factors. First, international professional standards confer prestige on scientists who work in pristine laboratories insulated from the world's poor. Second, the relatively strong research institutions which were inherited from colonial rule remain powerful in their new context because they serve export industries that remain economically and politically important. Political leaders have to make a sharp break with the past to develop subsistence or minor cash crops for the domestic market. Third, despite the urgency attached to domestic food production, export crops retain a strategic position in many economies because the international environment is biased toward trade (Lipton, 1985). External donors still advise aid recipients to pursue development through exports, further justifying continued emphasis on that part of the IATS geared to export crops (World Bank, 1984).

Contemporary Socio-Political Structures

The politics of the typical developing country are heavily influenced by patron-client networks. These harm the interests of both small-scale farmers and the IATS that are supposed to serve them.

Small-scale farmers in pre-industrial societies offset the risks of agriculture by investing heavily in personal relationships as a hedge against adversity. Not only may close ties with social equals, such as relatives and neighbors, provide help in times of drought or illness, but bonds to social or economic superiors can also be a source of assistance. For these latter unequal relationships, the recipient promises support in return for the help that he or she receives. This social dynamic lies at the root of the patron-client relationships that pervade pre-industrial societies and that dominate most of their political processes (Migdal, 1974).

Patron-client networks group together people of dissimilar interests; a patron must be advantaged in some area in order to have something to trade for the support of his or her clients. The resulting political processes are substantially different from those resulting from associational groupings, which bring together those who share a common interest (Scott, 1972). Associational politics lead to the direct representation of the interests of large or powerful groups in society. In contrast, patron-client politics mask the interests of the multitude, who are the clients, for they are represented in the system by advantaged patrons, whose personal interests differ from those of their followers.

Sometimes political systems are mixed, so that some groups are incorporated through patron-client networks while others are represented by associations. Such mixed systems generally work to the still greater disadvantage of small-scale farmers. In these kinds of systems, industrial elites, large-scale farmers and, sometimes, urban workers have associations to press their interests, while small-scale farmers have the expression of their interests inhibited by patron-client networks.

Associational politics tend to result in the creation of "public goods" for the most powerful groups, and policies that will serve the interests of the groups' memberships. Patron-client politics, on the other hand, tend to create "private goods" — discrete products and services, such as credit, that can be disaggregated and distributed to individuals through the networks (Bates, 1981). To the extent that patron-client systems produce any public goods at all they tend to be ones that benefit the elite group of patrons and which add to the personal wealth on which they can draw to maintain their networks. Patrons do not afford priority to the provision of technologies unless their distribution can be controlled in such a way that political support can be claimed in return.

The products of public sector research and extension are generally, though by no means invariably, public goods. They are intended to benefit groups of similarly situated people. Many new technologies — especially management practices and information — diffuse spontaneously once they are available, thereby denying profit to their creators. It is difficult to sell most agricultural research and extension services in such a way as to cover expenses, even though the benefit to society is usually significantly greater than the cost (Evenson and Kislev, 1975). This is why such a large proportion of agricultural research and extension services are supplied by the state, the usual vehicle for the provision of public goods.

To the extent that patrons are interested in research and extension systems, they are likely to focus on the distribution of credit and inputs or to treat them primarily as sources of employment for their clients, thus effectively turning them into purveyors of private goods with considerably diminished public benefit. Only those parts of the farming community that are organized along associational lines are likely to demand the kinds of research and extension that constitute public goods. In pre-industrial societies, associational representation of farm interests is usually confined to large, commercial enterprises.

Those parts of the IATS that serve the interests of small-scale producers consequently tend to be orphaned. The patron-client character of politics tends to dampen effective farmer demand for their services. As a result they are underfunded and become self-governing, with few external pressures on them. Of course, it is possible to have, alongside these orphans, other parts of the IATS that operate efficiently, creating and transmitting technologies to large-scale, commercial farmers. In fact, the articulate political demand by large enterprises for services geared to their needs often leads to resources being syphoned off from the part of the IATS which serve small-scale farmers. In pre-industrial societies, IATS generally will serve the interests of small farmers only to the extent that these interests happen to coincide with those of larger commercial enterprises.

Default Conditions

If there are few external demands for effective research and extension services for small-scale farmers in societies where patron-client politics prevail, what are the internal, default conditions that govern the operation of these services? The stylized "model" we present in this section is not a description of what always occurs, although something like it is found far too often. Instead, it illustrates the dangerous effects of inertia. In other words, we are asking: What conditions will prevail unless vigorous action is taken to change them?

Conditions for Research

First, scientists will produce for the agricultural research profession. They are conditioned to do so by education and training, which impart both their values and research methods. Scientists' orientation toward an audience of fellow professionals is also influenced by funding considerations and by professional rewards and incentives (Chambers and Jiggins, 1987). The hierarchical nature of scientific education conditions researchers to expect a high status in society. As experts, they see themselves as defining the issues which they believe ordinary citizens cannot fully articulate, and as solving society's problems in the modern laboratory.

Scientists' research methods traditionally exclude clients from formulating the problems and contributing to their solution. They are highly technical, and based on a reductionist approach involving a limited set of variables whose interrelationships are relatively easy to grasp, at least compared with the complexities facing resource-poor farmers. Conventional research methods are thus best applied to the conditions of relatively privileged producers, since they, like the scientists, have the capacity to control the natural environment (Busch and Lacy, 1983; Chambers and Jiggins, 1987).

Professional rewards and incentives drive scientists to produce scientific papers rather than serve rural constituents. Recognition and approval is sought from the research profession through publication, which scientists may also see as the principal means of disseminating their findings (Balaguru and Rajagopalan, 1986). Publication in international journals yields the highest recognition, leading perhaps to job offers from international research institutes and to new funding opportunities. Scientists are therefore motivated to select topics which will interest international research circles, even if they bear little relationship to the problems of resource-poor farmers in low-income countries. Where resources do not permit research for a wide scholarly audience, professionals strive to meet the requirements of senior officers in their institution instead of soliciting information and feedback from clients.

Regardless of whether or not approval is sought from the international scholarly community or from organizational directors, researchers will usually seek to maximize their personal comfort. One way to do this is to choose to carry out routine research on problems that are known to be solvable, thus avoiding the stresses caused by new and difficult research areas for which new approaches and methods have to be devised. A second way is to place more emphasis on laboratory projects than on field projects, because the working conditions of the former are more pleasant. Laboratory research conditions are also more subject to control, placing success within easier reach.

Scientists who work in laboratories are more likely to be located in cities, which offer a host of advantages. Since urban areas have resources which support conventional research, including libraries and equipment, scientists in such areas are in a stronger position for achieving recognition and promotion. Painstaking work carried out by scientists located in the countryside often escapes the attention of senior officials, thus undermining the morale of field staff. Another deterrent is the lack of amenities for researchers' families typical of remote research locations. Researchers who are posted on field stations are often oriented toward future prospects in urban areas. During their stay in the countryside, they emphasize on-station work rather than research in farmers' fields in order to minimize travel and avoid confronting complexity.

Scientists also seek to maximize their personal comfort by choosing to conduct research on major cash crops; the research stations for these crops are likely to have been established longer and to have better physical amenities than stations concerned with food crops and other commodities produced by resource-poor farmers. Farmers enjoy a socio-economic status that is not too far removed from that of researchers, facilitating client interaction and support. In addition, research on cash crops offers better prospects for employment in the more highly paid commercial sector.

Conditions for Extension

The "classic" extension organization in a patron-client society is poor at promoting integration with research. It is unresponsive to poorer farmers' legitimate needs and adds little to the relevance of the research system for the farming community as a whole. Its capacity to promote the adoption of innovations is likewise limited, unless the organization operates under strong hierarchical discipline.

Because the organization experiences very little politically effective demand for its services from small-scale producers, its work is highly biased toward the minority of farmers who are willing and able to register complaints higher up the hierarchy. The demands met in this way are often for inappropriate private goods, such as free labor on the recipient's "demonstration" crops, but they are few enough for extension agents not to be seriously inconvenienced by meeting them and so purchasing the recipient's silence (Leonard, 1977).

Professionalism is generally weaker among extension agents than among researchers, and the career rewards for adhering to professional standards are less compelling. Left to themselves, extension agents tend to work only with those farmers who are immediately responsive to their suggestions (Leonard, 1977; Thoden van Velzen, 1977), to give advice on the technical innovations they know best rather than trying to solve farmers' most important problems, and to concentrate on producers located near major thoroughfares rather than walking through the countryside to help more isolated farmers (Chambers, 1983). Such extension organizations do not seek new technical information from research institutes, for innovations might require change that would upset the existing equilibrium.

This "classic" extension system need not always exist in pre-industrial societies. There are certain situations in which powerful external incentives may arise to govern extension behaviour. The first occurs when harvests of food staples or export crops are so poor that the patrons/elites experience economic problems. The political leadership will then press extension to increase production. The incentives created in this situation can be substantial and lead to considerable improvements in performance, but they will be hierarchical in character and not directly responsive to farmers' concerns.

Extension in patron-client systems finds it difficult to respond to producers' expressions of their needs and does not serve as a significant channel for communicating farmers' most pressing problems and perspectives to the research system. Unfortunately, the more an organization stresses hierarchical pressure in order to improve performance, the more difficult it is to generate feedback. "Lowly" extension agents do not want to report problems to authoritarian and unsympathetic superiors.

The second exceptional circumstance in which external incentives may be created for extension occurs when larger, commercial producers are represented not by patrons but by associations, and thereby promote integration with research and more relevant recommendations, leading to wider adoption. However, such associations do not make extension more responsive to poorer farmers' needs. The latter will benefit only in so far as their conditions of production are similar to those of the commercial producers.

If the statistical distribution of farm sizes is unimodal, and larger farms are simply the more advantaged representatives of a single production system, such overlap in relevance is likely. If the distribution is bimodal, with two distinct groupings of large and small farms, common relevance is much more limited and small producers will suffer even greater neglect. In pre-industrial societies, the more the statistical distribution of farms conforms to the unimodal pattern, the less likely are associational patterns of politics among the farming elite.

Consequences for the IATS

Under default conditions, the part of the IATS that is oriented toward the rural majority tends to be fractured rather than integrated, and its components have no incentive to coalesce. Cooperation is impeded by differences in the nature of the tasks allocated to research and extension, which are associated with significant differences in status.

Researchers' work requires abstract analysis and specialized expertise and is quite specific compared with the diffuse activities of extension agents. The latter work with people amidst considerable uncertainty about goals and strategies. It is well known that success is both more likely and more widely recognized in organizations with specific, unambiguous goals than it is in those where goals and the means to achieving them are hazy and/or disputed (Israel, 1987; Thompson and Tuden, 1959). Under default conditions, the part of the IATS devoted to research appears more successful and enjoys higher status than that devoted to technology transfer.

Extension agents often feel they have little to gain but much to lose from collaboration with research personnel, and vice versa. Both would surrender autonomy, and interaction might yield no recognition from superiors. Thus researchers continue to seek approval from their professional colleagues, while extension agents try to satisfy the minimum requirements of their superiors.

Extension staff do not press researchers for more relevant research, nor do they respond to the concerns of farmers whose socio-economic status is lowlier than their own. The failure of farmers to adopt innovations is attributed to their conservatism and so poor adoption rates are not seen as an organizational problem.

Thus the default conditions associated with colonial and contemporary institutional and socio-political structures combine to maintain a status quo which saps the vitality of the institutions charged with agricultural research and technology transfer for the majority of rural producers.

Forces for Change

National Policy Makers

Does it matter that the public institutions for promoting research and development are weak? In agricultural revolutions of the distant past, and even in the more recent one in 18th and 19th century northwest Europe, the state played only a marginal role. Technological change resulted mainly from the inventions of society (Chambers and Mingay, 1966; Lipton, 1985).

When the most recent agricultural revolution unfolded in Asia in the 1960s, some economists interpreted it as a drama in which economic forces provided cues for public institutions — that is, the state's role was simply to follow the market (Binswanger and Ruttan, 1978; Hayami and Ruttan, 1985). Since then, attention has been drawn to the more active role of the state in technological and institutional change (de Janvry and Dethier, 1985; Johnston and Clark, 1982), suggesting a need to assign political factors a weight at least equivalent to that accorded to economic forces. The state can be a major independent factor shaping the nature and scope of agricultural change.

The commitment of various interests within state structures to developmental objectives is a crucial determinant of the policy environment within which change is to take place. Where commitment is lacking, fleeting or expressed formally but discounted in practice, resources will be dissipated in schemes which advance neither productivity nor welfare (Heaver and Israel, 1986). The scope for development may be severely limited if policy makers have close ties with groups such as landed elites, whose interests are inimical to development.

Such policy makers often show a lack of commitment to development, even in the face of severe problems. Shortages of food or capital may foster conditions which threaten a country's social, economic and political order, yet policy makers may still fail to take decisive action. Instead, the building of the state and the consolidation of central authority hold overriding priority.

These conditions are not immutable, however. Public institutions are not static monoliths, but form political systems embodying diverse interests (Heaver, 1982). At certain junctures, key people within the system who believe agricultural development should hold high priority may form informal coalitions to pursue that objective. Such coalitions can mobilize the support necessary to stop policies and programs aimed at the consolidation of power and start introducing practical measures to develop agriculture and hence the economy as a whole. Pressure for change appears more likely to develop and prevail in decentralized authority systems rather than in centralized ones, which suppress competing interests (Bacharach and Lawler, 1980).

History shows that the resolve of national leaders to develop the agricultural sector can be triggered by external challenges to national autonomy, which compound domestic problems. War, or the threat of it, undoubtedly plays an important part, as in the case of Britain's "Dig for Victory" campaign during the Second World War. The fact that external assistance is withheld, or offered on unacceptable terms, can also be important. In the case of Tokugawa Japan (1603-1867) and Maoist China, the need to mobilize indigenous resources was critical because external assistance was unavailable. (China had substantial assistance from the USSR in the early 1950s, but these ties were first weakened and then broken before China initiated major agricultural change.) In the case of India during the mid-1960s, assistance was offered on terms which policy makers found unacceptable. In postwar South Korea and Taiwan, policy makers felt that the amount of aid offered was insufficient in relation to the magnitude of the challenge posed by external forces (Rosen, 1985; Sims, 1988; Trimberger, 1978).

Policy makers who have mobilized domestic resources in an all-out drive for agricultural development have generally emphasized either institutional or technological forces of change, the choice being determined in part by the technology available and the physical environment in which it will be applied, and in part by the will of leaders to intervene in society and their capacity to do so. However, growth can be sustained only if institutional and technological strategies are integrated, and if the strategies are reinforced by measures which provide both the necessary infrastructure and a policy environment conducive to development.

Meiji Japan (1868-1912) offers the most striking illustration of state intervention emphasizing both institutional and technological development. Less striking, but still successful, examples are China, South Korea and Taiwan, where the mix of institutional and technological strategies differs from country to country and over time. India exemplifies a situation in which efforts to change institutions were highly selective, with greater reliance on technology as an agent of change. The contrasting examples of Japan, India and China merit further consideration.

Japan. The Meiji leaders were determined to restructure the institutions of a fragmented society. Their commitment arose from the threat of external domination, which led to a forced march toward industrialization financed by an increasingly productive agricultural sector.

Two factors enabled the governing elite to choose and implement its program for agricultural growth unimpeded by social and institutional constraints. First, the Meiji samurai leaders were not part of the dominant landed classes, having been divorced from the land for centuries (Trimberger, 1978). Thus they had unusual latitude for action, independent of the interests of this powerful socio-economic group. Second, the state's leaders had access to a bureaucracy of exceptional efficiency and capacity. Not only was the Meiji bureaucracy capable of directing change, but it was also quite responsive to the circumstances of its clients, with whom it was able to interact effectively.

The Meiji leadership recognized the need to enlist the cooperation of farmers in the search for the best local practices, and in their testing and dissemination. Farmers carried out location-specific research and served as itinerant instructors. Officials were expected to work with the farmers, and the collaboration between the two groups served to link the research and technology transfer functions, leading to the development and diffusion of innovations which were clearly applicable to users' circumstances. The adoption of innovations was facilitated by mechanisms for social control, and ensured through coercion (Hayami et al., 1982).

The Japanese model is not easily transferable. The factors that made it successful, such as social homogeneity and mechanisms for social control, are not widely found elsewhere in the developing world. It is particularly difficult to develop local institutions which can cooperate with their clientele and respond to its needs, while still channeling citizens' behavior so that it meets national priorities. Most state leaders therefore opt for a strategy which assigns science and technology a starring role, thereby allowing scope for considerable assistance from the international agricultural research community.

India. During the 1960s the international agricultural research community developed agricultural technology which was considerably more powerful than that available to the Meiji leaders, although its applicability was limited to restricted geographic areas. This technology was to prove sufficiently potent to bring food self-sufficiency to several countries, despite deficiencies in the local institutions charged with its promotion.

Technology is powerful in itself, but its capacity to transform rural society depends on how supportive the policy environment is, and on how the users of technology interact with the institutions which control the factors of production. India's success with improved wheat and rice varieties was associated with longstanding public support to infrastructure, including irrigation, roads and electrification, and to local government bodies and cooperative societies.

India's food crisis of the 1960s was exacerbated by the external pressures that had been exerted on its agricultural policy during the colonial era. In the fluid situation following the death of Prime Minister Jawaharlal Nehru, a coalition arose within the state structure to lobby for agricultural development. Its members recognized the importance of incentives to enlist the cooperation of key groups. Agricultural scientists won unprecedented recognition from policy makers. Research productivity was rewarded, rather than seniority or political loyalty. Farmers were given price incentives to adopt researchers' innovations.

Indian development emphasized the diffusion of new technology, but it did so in a top-down fashion. The ruling coalition was determined to achieve self-sufficiency in food grains. The results in terms of agricultural production were encouraging, but success was achieved through a hierarchical system that did not take into account the country's cultural, social and agro-ecological diversity. The institutions which could have facilitated interaction between researchers and farmers were short-circuited, because the green revolution technology was well adapted to the circumstances of farmers in the irrigated wheat-growing heartland.

India's single-minded pursuit of production increases left vast constituencies behind. It is increasingly clear that lagging groups and regions must not be ignored in the drive to modernize. A strategy such as India's, which focused disproportionate attention on technology, may not promote and certainly cannot sustain integration and responsiveness, which demand attention to institutions as well as technology; in such cases, relevance will be limited to geographical areas with conditions like those of research stations. To serve a wider rural clientele, policy makers must promote institutional change that will facilitate interaction between research and technology transfer staff, and support farmers' participation in technology development and diffusion. Such strategies were followed not only in Meiji Japan but also in Maoist China.

China. The Chinese case contrasts sharply with that of India. The Maoist leadership was committed not to agricultural growth alone but to a set of broader development aims for the rural sector, including improved health care, education and nutrition. This reflected both its populist orientation and its ties to a support base consisting of the rural poor. By 1980, as a result of state-led initiatives, poor Chinese generally had a better diet than their Indian counterparts, and the average life expectancy of 66-69 years compared favorably with India's 52 years. Despite the Chinese regime's emphasis on equity, agricultural production had shown creditable growth (Sen, 1986).

Development of the technology to drive increased production was not assigned solely to scientists and officials. Political leaders emphasized the value of practical knowledge over formal qualifications, and of mass participation even if this meant grossly inefficient use of resources. This approach prevented the officials and scientists from monopolizing technical knowledge and, as in Meiji Japan, encouraged a system of research and technology transfer that was integrated and responsive to a broad clientele which held center stage. The participation of farmers undoubtedly encouraged the development of relevant technology, thereby facilitating widespread adoption.

Since equity held such high priority, the Maoist leadership recognized earlier than many others that food policy encompassed more than just production (Gittinger et al., 1987). Success in the latter does not necessarily translate into increased food for the malnourished, as shown by many recent studies (for example, Pacey, 1986). Yet despite the impressive record of the Maoist leadership in reducing hunger and malnutrition, China's political system was not immune to forces which caused a marked shift in course, to the detriment of citizens whose access to food remained shaky (Sen, 1986). There are no formal mechanisms to ensure the system's continued responsiveness to society as a whole, let alone to disadvantaged groups and regions, which may fall further behind when policy makers' emphasis shifts to growth, as it did in China following Mao's death.

Donors

External aid agencies were instrumental in the development of many IATS in the postcolonial era. They could still do much to improve the integration and responsiveness of IATS. Yet certain structural characteristics restrict their ability to do so.

Most donor agencies are centralized operations with limited staff in the field; thus it is difficult for them to respond to local conditions. Moreover, donor agencies often encompass a wide range of interests, including geopolitical, commercial, bureaucratic and humanitarian concerns. Their decisions are often influenced by pressure to move large sums of money quickly. This encourages the widespread and rapid replication of programs developed previously, without first determining the modifications needed to suit local conditions (Duncan, 1986; Sussman, 1982; Tendler, 1975).

The pressures facing donors are revisited upon aid recipients, whose already overburdened administrative capacities are further taxed by the procedures involved in securing and managing aid funds. The infusion of large sums of money further diverts policy attention away from issues which are not readily solved through the infusion of large sums of money.

The gulf between research and technology transfer has only recently been identified as a problem by international donor agencies (World Bank, 1985), and their attempts to promote corrective measures have so far been limited. In the past, integration was hindered by the tendency of donors to focus assistance on either research or extension, but not both. During the 1950s, many development experts assumed that the technology developed in the West could simply be transferred to developing countries, where it could be popularized by institutions modelled along Western lines. They failed to see that two vital ingredients — an active clientele and effective site-specific research — were missing.

During the 1950s the politically motivated Point IV programs of the USA established extension agencies in many developing countries, especially in Latin America, and assigned the role of catalysts for development to officials analogous to American county agents. Since extension services were linked neither with indigenous research systems nor with active clientele groups, they soon degenerated into ritualistic exercises in the delivery of extension methodology (Rice, 1974).

The succeeding decade marked the more reflective but still myopic period of institution building, which established land grant universities in several developing countries. Institutions were still treated like jigsaw puzzles which could be broken up and re-assembled by experts along the lines of Western protypes. The fact that these prototypes had arisen in response to demand from clients and were sustained by their continued support was still ignored. The new land grant universities did not respond to rural constituents' articulated demands and often misjudged their unspoken needs. Where clients were marginalized, university staff sought support instead from professional bodies and an expanding international scientific establishment. The scientific community rewarded those who pursued "pure" research within narrowly defined disciplinary channels, far removed from the day-to-day concerns of field staff and farmers.

Research increasingly claimed priority over extension in international and domestic funding allocations as agricultural growth rates lagged but populations expanded. Economists' calculations of the relative contributions to productivity of agricultural research and extension indicated greater returns on investment in research, and reinforced perspectives which treated the two systems in isolation from one another (Byerlee, 1987).

The problems facing research were easier to resolve than those facing extension, and considerably less complex than the institutional changes required for effective integration of the IATS as a whole. Agricultural scientists could be given more financial incentives and new opportunities for recognition in national and international circles. Even farmers won respect in scholarly treatises when they responded to incentives to adopt new technology. In the meantime, extension agencies languished in the disillusion which had followed earlier efforts to transfer technology and institutions.

The focus on research is epitomized by the establishment of the international agricultural research centers (IARCs) of the Consultative Group on International Agricultural Research (CGIAR). These centers have emerged as a major source of new agricultural technology and, through farming systems research (FSR), have begun to address the problems of research relevance and responsiveness. The shortcomings of extension systems have been the focus of the Training and Visit (T and V) system developed by the World Bank.

We will now turn to the positive effects of these three initiatives launched and supported by donor agencies, but note that the tendency to treat research and extension as separate systems remains a problem, as does the prescription of universal rather than nationally derived solutions.

International agricultural research centers. The new emphasis on agricultural research appeared to be justified by the spectacular successes of the green revolution. The rapid diffusion of new high-yielding varieties of wheat and rice suggested, at least initially, that attention to formal extension systems might not be necessary when researchers could develop technology offering high returns and policy makers could facilitate its adoption through price incentives and supporting inputs.

As the green revolution subsided, resource management became more complex in the favored areas where the agricultural transformation had taken place. Broad prescriptions lost their usefulness as farmers pondered site-specific questions. Meanwhile, the more intractable problems of farmers in Africa and other resource-poor regions became a subject of increasing concern.

In the late 1970s the IARCs began to widen their focus and to serve the vast constituencies who had not participated in the green revolution. New centers were established in Africa. Scientists paid increased attention to the food staples grown by resource-poor farmers, and included employment opportunities as a consideration in their research.

The IARCs deserve major credit for the increased scientific awareness of the needs of small-scale farmers and the requirements of crops and livestock species which had previously been ignored. As international bodies which are not tied to specific national objectives, the IARCs are free to set research priorities which address the needs of clients excluded by the elite coalitions prominent in national policy-making processes (Lipton, 1985). The significant shift in their concerns toward resource-poor farmers has begun to influence the agendas of national research systems.

The IARCs are oriented neither to the frontiers of biological research nor to the conditions of intensive agriculture (as, for example, are American land grant universities). They do not pursue research topics which are commercially profitable, as do corporate laboratories. The IARCs' professional ethic is dedicated to the creation of technologies relevant to extensive production by smallholders and pastoralists.

The impact of the CGIAR system extends well beyond its technological achievements. Probably as important has been the magnetic pull it has exerted on researchers in national systems, creating a new definition of professional excellence that stresses relevance to small-scale producers, providing training for and interaction with national researchers, and improving the global information systems that now link previously isolated researchers in different regions, countries and continents. The IARCs offer incentives in the form of publication outlets and professional recognition for those who pursue research for a non-professional audience (Anderson, 1985). Under their tutelage, professionalism becomes a much more positive force than it would be otherwise.

The extent to which the IARCs are able to redirect research priorities in national systems toward greater relevance varies. The links between IARCs and national institutions may be weak or strong. Sometimes the very superiority of the resources available to the IARCs provokes a defensive response in the national system. Local scientists then try to keep their distance in order to protect themselves from what they see as a "neo-colonialist" intellectual take-over. This reaction testifies to the powerful influence that the IARCs can have.

The impact of the IARCs in each country is affected by the weight of various domestic constituencies and by the priorities of national leaders, as well as the economic and physical resource base. If the political leadership reflects the interests of a powerful minority of commercial growers, the IARCs will have only a limited impact on the capacity of the national IATS to serve the resource-poor majority. On the other hand, political leaders may wish to serve the rural majority, but lack the necessary resources to do so. In such cases, IARCs can make an important contribution to the relevance and responsiveness of research.

Since the 1970s, the IARCs have tried to promote responsiveness by broadening scientists' perspectives through multidisciplinary research, including the social sciences. That this development is regarded with ambivalence even within the IARCs is clear from the pressure being experienced within the system to restore the previous pattern in which the IARCs specialized in the biological sciences and left disciplines such as the social sciences to national systems (CGIAR, 1985). Further, the agricultural scientists who predominate in the IARCs were trained to think in specialized disciplinary channels which are too narrow to encompass the world of the average resource-poor farmer, let alone that of the women who constitute the majority of the world's food producers (Jaquette, 1985). In other words, although the IARCs as institutions have developed a new professional ethic in research, some of the old habits and attitudes linger on in individual scientists.

The IARCs generally have a limited capacity to respond to purely local needs. Their links with national technology transfer organizations are often tenuous—and it is physically impossible for their limited staff to cover vast expanses of ecologically diverse terrain. Thus, with regard to integration, the IARCs do not yet bridge the chasm between agricultural research and technology transfer. Perhaps this will change as donor pressures lead them to do more field trials with national extension staff. This trend is not likely to occur spontaneously, arising from within their internal, professional incentive system—a further indication of the importance of donor perceptions of problems and funding priorities.

Although the CGIAR centers now try to respond to the needs of resource-poor producers, it is proving difficult to design technology suited to their needs. FSR, to which we will turn next, represents a major step in that direction.

Farming systems research. The FSR approach seeks to increase relevance and responsiveness through multidisciplinary, location-specific field research emphasizing the solving of practical problems. FSR does not necessarily focus on the rural poor, and thus does not inevitably clash with the interests of rural elites, who exert an influence in some domestic policy making systems. FSR may nevertheless be "disarticulated" from many IATS because its approach differs significantly from the established practices of agricultural scientists, and integrative mechanisms are often weak (Marcotte and Swanson, 1987).

FSR emphasizes the adaptive research which is essential if the needs of the rural population are to be served. The capacity of its practitioners to serve this large constituency may be diminished by their distance from the professional mainstream which, as we have seen, confers rewards and prestige on scientists who conduct strategic and applied research on well equipped stations. As "deviants", FSR practitioners enjoy less prestige than other national and international scientists, on whose priorities they cannot exert much influence. Since FSR often lacks strong indigenous political backing, it is vulnerable to charges that it embodies external donor interests rather than local ones.

Some observers have categorized FSR as yet another top-down approach to the transfer of technology, albeit a more cunningly disguised one. The acid test of FSR is its responsiveness. As it has been practised so far, FSR has not usually provided for clients' direct participation in setting the research agenda. Only recently have its proponents begun to promote methods that change the roles of scientists and farmers in the process of technology generation. Nor are the incentives for such a change very strong even now (Chambers and Jiggins, 1987).

At least FSR does not preclude participation — and many FSR teams are indeed responsive to the needs of farmers, which they identify through diagnostic research, a process similar to the market research carried out by the private sector. They can and do help to persuade biological scientists of the legitimacy of farmers' views, thereby promoting the relevance of research. Because of their use of sample surveys, the social scientists at work in FSR teams can be more aware than other scientists of the needs of resource-poor farmers. Thus they can also improve the responsiveness of research.

Social scientists need to be well integrated within interdisciplinary teams, and not simply serve as window dressing, if FSR is to lead to greater research relevance and responsiveness. One of the problems is that biological researchers have higher status and are usually more senior. It may be unrealistic to expect them to allow junior social scientists to set their research priorities unless the latter have substantial support from further up the hierarchy.

A major criticism leveled against FSR is that it tends to by-pass formal extension services in its efforts to bridge the gap between researchers and farmers. Extension needs to "co-evolve" with research as the latter becomes more relevant to its clients. It must assume a more active role in the process of technological change, otherwise it will become increasingly irrelevant (Johnson and Claar, 1986). This is unlikely to happen automatically. It will occur only if FSR redefines its mission to incorporate extension in its diagnostic research and on-farm trials, and encourages researchers to learn from extension staff as well as from farmers.

In sum, FSR's success in promoting relevant and responsive research depends on the strength of its incentives to do so, and on the nature of the interaction of its scientists with their colleagues in research, extension and the farming community. The incentives provided to FSR teams must be weighed against the rewards offered to the professional mainstream in order to predict how individual researchers will respond. FSR researchers are more likely to be concerned with adoption than their station-bound colleagues because they encounter non-adopters as well as adopters, thereby receiving potentially useful feedback. The major problem that remains to be addressed is the integration of FSR activities with those of station-based scientists and extension agents.

The Training and Visit system. The third major initiative undertaken by donors focuses on extension. The T and V system is more concerned with the links between research and extension than is FSR. It does not address the problem of responsiveness, but pays greater heed to adoption than does FSR. A major weakness of the system is its assumption that research is relevant, and that its results need only be delivered to farmers.

The T and V system attempts to make extension activities more specific, and thereby facilitate interaction with researchers, whose work is inherently more precise (Israel, 1987). The system's key contribution is its establishment of formal links between research and extension through subject-matter specialists, who serve as mediators and participate in regular training sessions.

The efficacy of formal links is shaped by professional values, the strength of incentives to cooperate and the extent of support from local, national and donor interest groups. If senior research staff wish to develop better relations with their rural clients, they may welcome the chance to train extension personnel and become involved in field activities. But where researchers are oriented toward a professional audience, they will view training sessions and missions to the field as burdensome tasks which deflect energy from the preparation of publications. In the first case, extension staff will probably appreciate the opportunity to enhance their status, whereas in the second they will resent the claims on their time and the didactic approach of their trainers. In the latter case, integration will be purely a formality, tailored to meet donors' requirements.

Like FSR, the T and V system is often seen as representing external interests and priorities more than domestic ones. If domestic policy makers are not fully committed to it, T and V practices will be diluted and modified. Actual performance is determined at the field level, by staff who hold decisive informal power in the realm of policy implementation (Wagemans, 1987).

The incentives for senior officials in departments of agriculture to comply with the organizational demands of the T and V system are frequently both visible and compelling. The system's top-down approach is easily accommodated in hierarchical bureaucracies whose leaders condone such an approach. Policy makers often appreciate new mechanisms for holding field staff accountable for performance. Yet the incentives and conditions needed to fulfill T and V requirements at the field level, where they are most needed, are often quite weak.

The T and V system's success depends on the vitality of adaptive research. Where such research is deficient, extension staff may be held accountable by their superiors for the regular delivery of information which is of little value to clients. In such cases, extension staff soon lose both the respect of their clients and their own morale. The ability of subject-matter specialists and field staff to convey feedback to scientists and influence their research priorities and design is therefore crucial. Occasionally, extension agents are able to take the initiative in applying information creatively, according to the varying needs of their clientele (Wijeratne, pers. comm.).

As we noted earlier, hierarchical systems such as T and V hinder feedback from subordinate staff and clients. While the T and V system clarifies what extension agents are supposed to do, it does not address their major grievances, including low pay and inadequate logistical support in the field. The rigid schedule of training and visits facilitates monitoring and evaluation, but does not take constraints at the local level into consideration.

The T and V system is most successful where producers are relatively advantaged in terms of resources and agro-ecological conditions, and operate farms where conditions are relatively homogenous and broadly similar to those of research stations. These circumstances prevail in the north Indian state of Haryana, where the system is said to have shown positive results (Feder et al., 1985). Farmers in Haryana enjoy access to agricultural inputs and credit thanks to the extensive involvement of both the public and the private sectors.

The T and V system has trouble responding to farmers who do not share such advantages. Resource-poor farmers face extremely variable conditions in which the information and technology of broad applicability typically disseminated through the T and V system have little value. Even in favored regions such as north India, many farmers now need a more detailed understanding of technological requirements than is afforded by the general prescriptions of the T and V system (Byerlee, 1987). Resource-poor farmers often want credit and inputs much more than they want advice that does not help them increase production without incurring additional costs. Hence, the T and V system is weak in terms of relevance and responsiveness.

Evidence on the extent to which the T and V system has promoted adoption is still limited and mixed. Feder et al (1985) reported high adoption in Haryana, but Khan et al (1984) found that there had been no significant impact in Pakistan's Punjab. Byerlee (1987) noted mixed results. It appears that adoption is high where farmers and researchers face similar environmental conditions, provided that adaptive research is adequate. In any event, increased attention to the criterion of adoption represents a positive development.

Extension staff suffer losses as well as gains from the attempts of the T and V system to streamline their institutions and upgrade their training and performance. They may lose autonomy and informal income as a result of their narrowed mandate, limiting their tasks to advisory functions. In addition, while the responsibilities of field staff may be reduced in theory, they may remain broad in practice because policy makers persist in assigning them additional responsibilities which further officials' own interests at the local level.

In sum, the T and V system addresses the problem of integration through the extension system. Its impact on relevance is limited since its concern is mainly with extension and its hierarchy inhibits feedback. It does not promote responsiveness. Since it does not expand the supply of relevant research findings, its impact on adoption will be strong only if such findings are already available. But at least it directs attention to this performance criterion.

Farmers' Organizations

There is considerable evidence to suggest that the strength and character of farmers' organizations are the single most important determinant of IATS effectiveness. In some political systems state leaders have promoted these organizations; in others, they have developed spontaneously, in the face of indifference or even resistance on the part of the state. Wherever they exist they seem to be important for the quality of agricultural research and extension.

The importance of farmers' organizations is that they directly represent the users of agricultural research. To the extent that they are effective in transmitting the needs of their members, they will demand relevant research, press for the integration of research and technology transfer and, as a consequence, promote adoption to a greater degree than do any other actors in the political or bureaucratic system. The managers of research stations and extension organizations frequently regard the interventions of farmers' organizations as technically ill-informed, short-sighted and generally a nuisance. Yet these organizations are vital in keeping research and extension on their toes, exerting pressure on them to integrate, and providing political support for better funding and policies for agriculture as a whole.

The history of agricultural development in the USA and Japan is often cited to illustrate the ability of governments to heed demands from their rural clients to improve agricultural services (Binswanger and Ruttan, 1978). In both cases, the government helped stimulate demand for innovations by organizing farmers' associations (McConnell, 1959). These bodies were seen as crucial in mobilizing mass participation in the quest for progress. The benefits of experimentation by farmers and the exchange of knowledge within the farming community had been illustrated by the extension systems which farmers had organized themselves in 18th century Britain. Active farmer involvement in the generation and diffusion of technology abolishes the artificial distinctions between these processes that are so often maintained by formal organizations.

Relatively few contemporary governments in developing countries have actively promoted farmers' organizations (Taiwan is one example). Unfortunately, most policy makers believe that the political risks incurred by encouraging farmers to form organizations outweigh the economic benefits. Governments lack the capacity to restrict the focus of farmers' associations to strictly technical issues, and are unable or unwilling to address broader ones.

Nonetheless, the initiative to organize farmers' associations need not rest with public officials, who neither work with the soil nor bear the risks of doing so. Demand for improved agricultural services has usually been pressed most effectively by highly commercialized producers, who have political and economic power. As producers of cash and export crops, they command the leverage needed to direct the attention of the IATS to their concerns, for they can provide powerful political support in return. The urgent need for foreign exchange, so often a major government preoccupation, combines with the needs expresssed by the interest groups representing commercial agriculture to produce strong pressures for relevant research and the adoption of its results.

Farmers' organizations tend to be more influential when their members are relatively well educated and enjoy access to resources which help them absorb the risks of innovation. The power of farmers' organizations is considerably influenced by whether or not they fund research, since funding sources influence the choice of research priorities. In societies where farmers contribute resources to agricultural research — as they do in the Netherlands, for example — scientists have clear incentives to address their problems, but in most developing countries funding is usually provided by government and commercial interests. Both the latter generally give priority to export crops, a rational choice in an environment overshadowed by foreign exchange problems.

As noted earlier, peasant producers tend to organize along patron-client rather than associational lines. However, the increasing commercialization of production is causing a gradual weakening of patron-client ties. New organizations are emerging in the countryside. These appear to be associational while still relying heavily on patron-client "templates" to structure internal relationships. Where these organizations are specifically agricultural, their leaders tend to be larger farmers, who can act as patrons. Nonetheless, the needs of small-scale producers may also be represented if the distribution of land resources is relatively equitable (or unimodal), for then the more advantaged farmers share similar production conditions to those of the less advantaged.

Where bimodal distributions of farm size occur, farmers' organizations (or less structured lobbies) tend to represent the interests of the larger enterprises, with smaller farmers remaining on the sidelines, waiting for innovations suited to their circumstances to "trickle down". To the extent that there are significant differences in the production systems of the two groups, the small-scale farmers will become increasingly disadvantaged. The factor endowments of large- and small-scale farmers are usually such that the large-scale ones benefit by capital-intensive new technologies while the small-scale ones gain from more labor-intensive ones. The political strength of the large-scale producers is likely to lead research to stress capital-intensive technologies, and small enterprises will become marginalized as a result.

We need to know more about the effects of benefits accruing to large-scale producers on the interests of small-scale producers in the technology development and dissemination process. To the extent that large-scale farmers dominate farmers' organizations they will both increase the political effectiveness of these organizations and bias the IATS toward meeting their needs. For small-scale producers there is a trade-off here: more and better research and extension may be obtained, but these may be less responsive to their needs. The question is where, on balance, this trade-off becomes disadvantageous to them. If both groups of producers are growing the same crops and using some, at least, of the same technology, resource-poor farmers may gain considerably more benefit from the political ability of the large owners to lobby for agricultural interests than they lose in bias of the system against their particular needs. We do not know.

The Colombian Coffee Growers' Federation is a case in point (Kaimo-witz, 1988). The federation was neither promoted by the state nor had any explicit political functions. It grew out of the efforts of large-scale coffee producers to organize themselves. Through their influence it was granted monopsony powers by the state, which taxed the coffee industry in order to provide it with services. The federation now provides research and exten-sion services which score high on integration, relevance and adoption, and from which small-scale producers have clearly benefited to some extent. The services do less well on responsiveness to the needs of the poorest coffee farmers, however, being biased against research on their particular prob-lems.

Diversity militates against the development of commodity-specific farm-ers' organizations. The typical low-income country is culturally and ecol-ogically heterogeneous, and farmers' organizations show marked regional variation. If a crop important in the national economy is produced on small plots throughout the country, farmers are unlikely to organize because the crop, though widely grown, remains a minor one for each farmer. If it is cultivated on a larger scale in a particular region only, producers there are more likely to organize (Piñeiro and Trigo, 1983).

The political power of farmers and their organizations varies considera-bly according to the crop(s) involved. Apart from export crops of strategic national importance, some food crops are widely consumed and marketed domestically, while others are not. It is the producers of subsistence crops consumed mainly on the farm who are least organized and represented in the policy-making process. These producers live on the periphery of society and its formal agricultural knowledge systems. The research bodies charged with subsistence crops are generally centralized and therefore weak at the local level, where adaptive research is most needed. Under these circum-stances, external assistance may be needed to help create farmers' organi-zations to work with researchers and extension agents.

Farmers' organizations which express the interests of wealthier farmers may simply seek advantages for this group at the expense of resource-poor farmers. In general, however, elite farmers are more interested in techno-logical development than are disadvantaged farmers. Organizations repre-senting resource-poor farmers tend to focus on issues such as land tenure. Such organizations may indirectly improve research and extension in the long term, if elite coalitions come to see resource-poor farmers as a constitu-ency whose demands require some response, even if not the one that is asked for. If the elite rules out the redistribution of land, for example, it might decide to pursue productivity improvements on smallholdings instead.

On balance, it seems likely that any type of farmer organization is better for the effectiveness of the IATS than none at all.

Commercial Firms

As already noted, commercial interests alone are unlikely to meet all a country's needs in technology development and transfer. Many needs are best met through public goods from which it is difficult to generate a profit. Some technologies have the attributes of private goods, however, and private enterprise will grow up around them if it is permitted. As we shall see shortly, veterinary services and farm machinery are two such technologies.

For those goods and services it markets, private enterprise has very positive effects on integration, relevance and adoption, for these criteria are closely related to a firm's profits. In this respect there are substantial benefits to be derived from private sector research and development. Private enterprise can also put pressure on the public sector to perform better.

Private enterprise poses a danger for responsiveness, for it will focus on the needs of wealthier producers. It will be oriented toward the development of discrete products whose distribution can be controlled and which can therefore be marketed for profit — that is, private goods. Commercial firms will not be interested in good management practices which farmers can communicate to each other by word of mouth, nor in improved seeds which farmers can breed themselves — that is, in public goods. These biases are not a problem as long as commercial interests do not distort public sector priorities to suit their own agendas.

Veterinary services. The veterinary field differs substantially from other areas of agricultural research and extension because the distribution of most of its innovations can be controlled and are therefore private goods. Curative veterinary medicine is also the area of agricultural services that has the highest producer demand: a farmer with a sick animal is extremely conscious of his or her need for assistance. As a result, in countries with a sizeable livestock sector, veterinary research and extension generally enjoy much wider political support and grass roots demand than do most crop-oriented services (Leonard, 1987). The most closely comparable area of crop production is the treatment of plant diseases and pests, which also enjoys high demand.

The widespread awareness of veterinary problems and the broad base of political support for their redress makes veterinary medicine one area in which the involvement of farmers' organizations is likely to be positive for all four performance criteria. Such involvement is likely to be beneficial even if the organizations are dominated by elites. Elite and resource-poor farmers in the same area tend to raise the same animal species, susceptible to the same diseases. If anything, the crossbred animals kept by elites are more susceptible to disease than are the traditional breeds raised by the less well-advantaged farmers and pastoralists. The increased risk of disease discourages such producers from adopting improved breeds. Thus the benefits of developments in the veterinary field are much more likely to be widely shared than is the case with crops.

The involvement of commercial pharmaceutical companies is likely to be positive in terms of integration, relevance and adoption. It may also be beneficial in terms of responsiveness, although to a lesser degree because of the limited purchasing power of the poor. The increased availability of information and products resulting from commercial activity has a broadly positive effect. However, businesses have little interest in promoting improved management practices that reduce the incidence of disease, and government extension efforts are normally necessary in this area.

Government can have two opposite effects on responsiveness, depending upon how it is involved. If the government service reaches a broad clientele, it increases demand from subsistence producers and hence increases commercial interest in manufacturing products which are tailored to this group. If, on the other hand, the government service is relatively ineffective, it may instead stifle demand from poorer producers by "skimming" the market. A weak or undisciplined public system tends to subsidize services for the more commercial and politically influential livestock owners, making it uneconomic for private enterprise to meet the remaining demand. In this case, the effect is to depress commercial pharmaceutical interest in manufacturing products for resource-poor farmers (Leonard, 1987).

Farm mechanization. In contrast to veterinary medicine, farm mechanization is an area in which the needs of elite and subsistence farmers are likely to be radically different, for the two groups are distinguished above all else by the capital they own.

If elite farmer groups are strong, mechanization is likely to be detrimental to the poor, both as subsistence producers and as farm laborers. The technologies developed will depress demand for farm labor and reduce the competitive advantage of farms with more labor than capital. Poorer producers tend to be better served instead by technologies that help them to break labor bottlenecks or increase the efficiency of their labor, rather than replace it.

Commercial involvement in the development and extension of farm mechanization will promote integration, adoption and relevance, but it will be negative for responsiveness. Nonetheless, it is possible for private enterprise to benefit poorer producers through mechanization, provided they have some cash to invest and represent a large market, as happened in Japan after the Second World War, when machines such as roto-tillers were widely sold to smallholders.

Government efforts in farm mechanization are not necessarily superior to those of the commercial sector. The problems of integration and relevance are more severe with regard to mechanization than they are in other areas. The development of appropriate technology depends on bringing together not only the perspectives of the farmer and the researcher but also that of the manufacturer. Very few government efforts have been known to surmount this barrier without the involvement of private firms.

Summary and Conclusions

In developing countries, historical and contemporary factors led to the establishment of agricultural research institutions oriented toward an international market and an international professional audience. There is a gap between research and technology transfer institutions, and an even wider breach exists between technology transfer institutions and the majority of agricultural producers in developing countries. Development can take place only if this situation changes, whereby institutions, technology and client groups coalesce in the search for the new agricultural technology which is urgently needed as the next century approaches.

The effectiveness of IATS can be assessed according to four criteria: integration, relevance, responsiveness and adoption. Four groups of people can force technology development and transfer systems in these directions: national policy makers, external donors, farmers' organizations and commercial firms.

There are only a few examples of effective state intervention combining technological and institutional change. The state has seldom managed to elicit the broad participation of its rural people in transforming an extensive tradition-based agricultural system into an intensive science-based one. Yet we do not rule out the possibility that coalitions will arise within state structures and prevail over the powerful default conditions that lead to inertia.

Forces related to the international environment have a mixed impact on national IATS. The international economic system combines with the international scientific community to offer compelling incentives to national professionals to conduct research on key export crops. Donor agencies have hindered integration by focusing attention on either agricultural research or extension, rather than on both together. Yet they have made significant strides toward responsiveness in some cases and relevance in others, notably by launching the IARC system and the FSR movement.

The pressures and competition provided by commercial firms are positive in the domains where significant profits can be captured, but if market demand alone drives the system, large constituencies will be slighted. Much will depend, in the future, on the recognition of the poor as a market, and the steps taken to give them purchasing power. At present, the private sector is still largely unresponsive to the needs of poorer producers, and neglects innovations whose benefits are primarily collective (public goods).

The most powerful demand for integration is likely to come from those who are, or should be, at the center of technology development and dissemination — farmers. The knowledge of resource-poor farmers is seldom tapped by external actors, and a complex web of social, economic and political factors inhibits the farmers' articulation of their interests. Among them is the traditional involvement of peasants in patron-client relationships, which prevents the expression of their collective as opposed to their individual interests. Agricultural research and extension are particularly negatively affected by such relationships. Although the commercialization of agriculture is weakening these patron-client ties, they still pervade most farmers' organizations. Associations representing the collective interests of peasants are now developing, but their purposes are sometimes subverted.

Resource-poor farmers need to be helped to recognize their collective interests as producers, over and above the socio-economic and personal differences which divide them at present. Donor agencies and national policy makers should encourage them to organize themselves so as to interact more effectively with research and extension. The participation and support of producers generates strong pressures for integration, relevance, responsiveness and adoption.

The first step in improving the performance of IATS in the developing world is to recognize the very real social and political shackles that bind them. Only then can one move intelligently to loosen them as opportunities arise, rather than flaying at them mindlessly or passively accepting their constraints.

Acknowledgments

We wish to thank David Kaimowitz, Niels Röling, Louk Box, Mahinda Wijerathe, Marie-Helene Collion, N'Guetta Bosso and Manon Kleinveld for their assistance during the preparation of this paper.

References

Anderson, J.R. "International agricultural research centers: Achievements and potential, Part 3." Washington, D.C.: CGIAR, 1985.

Bacharach, S.B., and Lawler, E.L. *Power and Politics in Organizations.* San Francisco: Jossey Bass, 1980.

Balaguru, T., and Rajagopalan, M. "The management of agricultural research projects in India, Part 2." *Agricultural Administration and Extension* (vol 23, no 1, 1986).

Bates, R. *Markets and States in Tropical Africa: The Political Basis of Agricultural Politics.* Berkeley: University of California, 1981.

Binswanger, H., and Ruttan, V. (eds). *Induced Innovation.* Baltimore: Johns Hopkins University Press, 1978.

Busch, L., and Lacy, W.B. *Science, Agriculture and the Politics of Research.* Boulder: Westview Press, 1983.

Byerlee, D. "Maintaining the momentum in post-green revolution agriculture: A micro-level perspective from Asia." MSU International Development Paper. East Lansing: Michigan State University, 1987.

Consultative Group on International Agricultural Research. "TAC review of CGIAR priorities and future strategies." Washington D.C.: CGIAR 1985.

Chambers, J.D., and Mingay, G.E. *The Agricultural Revolution 1750-1880.* London: B.T. Batsford, 1966.

Chambers, R. *Rural Development: Putting the Last First.* London: Longman, 1983.

Chambers, R., and Jiggins, J. "Agricultural research for resource-poor farmers: Part I." *Agricultural Administration and Extension* (vol 27, no 1,1987).

de Janvry, A., and Dethier, J-J. "Technological innovation in agriculture." CGIAR Study Paper. Washington D.C.: CGIAR, 1985.

Duncan, A. "The effectiveness of aid to Kenya: A case study." *IDS Bulletin* (vol 17, no 2, 1986).

Evenson, R.E., and Kislev, Y. *Agricultural Research and Productivity.* New Haven: Yale University Press, 1975.

Feder, G., Lau, L.J., and Slade, R.H. "The impact of agricultural extension: A case study of the training and visit system in Haryana, India." Working Paper. Washington D.C.: World Bank, 1985.

Gittinger, J.P., Leslie, J., and Hoisington, C. *Food Policy: Integrating Supply, Distribution and Consumption.* Baltimore: Johns Hopkins University Press, 1987.

Hayami, Y., and Ruttan, V. *Agricultural Development: An International Perspective.* Baltimore: Johns Hopkins University Press, 1985.

Hayami, Y., Ruttan, V., and Southworth, H. *Agricultural Growth in Japan, Taiwan and the Philippines.* Honolulu: University of Hawaii Press, 1982.

Heaver, R. "Bureaucratic politics and incentives in the management of rural development." Working Paper. Washington, D.C.: World Bank, 1982.

Heaver, R., and Israel, A. "Country commitment to development projects." Discussion Paper. Washington D.C.: World Bank, 1986.

Israel, A. *Institutional Development: Incentives to Performance.* Baltimore: Johns Hopkins University Press, 1987.

Jaquette, J. "Women, population and food: An overview of the issues." *Women and Food Production in Developing Countries.* Monson, J., and Kalb, M. (eds). Los Angeles: University of California, 1985.

Johnson, S.H., and Claar, J.B. "FSR/E: Shifting the intersection between research and extension." *Agricultural Administration* (vol 21, no 2, 1986).

72   *Sims and Leonard*

Johnston, B., and Clark, W. *Redesigning Rural Development: A Strategic Perspective.* Baltimore: Johns Hopkins University Press, 1982.
Kaimowitz, D. "Linking research and technology transfer in the development of improved technologies for small commercial coffee producers in Colombia." ISNAR Staff Note. The Hague: ISNAR, 1988.
Khan, M., Sharif, M., and Sarwar, M. "Monitoring and evaluation of Training and Visit system of agricultural extension in Punjab, Pakistan." Lahore: Punjab Economic Research Institute, 1984.
Leonard, D. "The supply of veterinary services: Kenyan lessons." *Agricultural Administration and Extension* (vol 26, no 4, 1987).
Leonard, D.K. *Reaching the Peasant Farmer: Organization Theory and Practice in Kenya.* Chicago: University of Chicago Press, 1977.
Lipton, M. "Modern varieties, international agricultural research and the poor." CGIAR Study Paper. Washington D.C.: CGIAR, 1985.
Marcotte, P., and Swanson, L. "The disarticulation of farming systems research with national agricultural systems: Bringing FSR back in." *Agricultural Administration and Extension* (vol 27, no 2, 1987).
McConnell, G. *The Decline of Agrarian Democracy.* Berkeley: University of California Press, 1959.
Migdal, J. *Peasants, Politics and Revolution: Pressures toward Political and Social Change in the Third World.* Princeton: Princeton University Press, 1974.
Pacey, A. *Agricultural Development and Nutrition.* Boulder: Westview Press, 1986.
Piñeiro, M., and Trigo, E. (eds). *Technical Change and Social Conflict in Agriculture: Latin American Perspectives.* Boulder: Westview Press, 1983.
Rice, E.B. *Extension in the Andes.* Cambridge, Massachusetts.: MIT Press, 1974.
Rosen, G. *Western Economists and Eastern Societies.* Baltimore: Johns Hopkins University Press, 1985.
Scott, J.C. "Patron-client politics and political change." *American Political Science Review* (March, 1972)
Sen, A. "Development: Which way now?" *Development Studies: Critique and Renewal.* Apthorpe, R., and Krahl, A. (eds). Leiden: E.J. Brill, 1986.
Sims, H. *Political Regimes, Public Policy and Economic Development.* New Delhi: Sage Publications, 1988.
Sussman, G. *The Challenge of Integrated Rural Development in India.* Boulder: Westview Press, 1982.
Tendler, J. *Inside Foreign Aid.* Baltimore: Johns Hopkins University Press, 1975.
Thoden van Velzen, H.U.E. "Staff, kulaks and peasants." *Government and Rural Development in East Africa.* Cliffe, L., Coleman, J.S., and Doornbos, M.R. (eds). The Hague: Martinus Nijhoff, 1977.
Thompson, J.D., and Tuden, A. "Strategies, structures, and processes of organizational decision." *Comparative Studies in Administration.* Thompson, J.S. et al. (eds). Pittsburgh: University of Pittsburgh Press, 1959.
Trimberger, E.K. *Revolution from Above: Military Bureaucrats and Development in Japan, Turkey, Egypt and Peru.* Brunswick: Transaction Books, 1978.
von Blanckenburg, P. "Agricultural extension systems in some African and Asian countries." Paper on Economic and Social Development. Rome: FAO, 1984.

Wagemans, M.C.H. "Voor de verandering." Ph.D thesis. Wageningen: Agricultural University of Wageningen, 1987.

World Bank. "The World Bank's agricultural research and extension: An evaluation of World Bank experience." Washington, D.C.: World Bank, 1985.

World Bank. "Toward sustained development in sub-Saharan Africa." Washington D.C.: World Bank, 1984.

3

The Effect of Changes in State Policy and Organization on Agricultural Research and Extension Links: A Latin American Perspective

Roberto Martínez Nogueira

One of the most widely debated issues in the field of agricultural research policy, organization and management is how to establish effective links between research and extension. The degree to which these two activities should be integrated, and the nature of that integration, has been examined from many perspectives and within many contexts.

Broadly, the problems related to creating effective research-extension links are similar to other integration problems within the public sector. However, because of the nature of research and extension activities, and the way in which they are incorporated into the state's organizational structure, they have specific linkage problems.

The purpose of this paper is to analyze the correspondence between the changing role of the state and the organization of government research and extension, and to examine the effect of this correspondence on the links between research and extension. Thus, the scientific and technological aspects of the links (which relate to the translation of contributions from different disciplines into practical and relevant technologies that can be transferred to the farmer) are not a core feature of the analysis. The emphasis is placed instead on research-extension links in the political/institutional and organizational/managerial spheres. The analysis adopts an historical approach, within the context of Latin America.

The paper is divided into three parts. In the first part, the analytical framework is presented and its central issues defined. This is followed by an historical review of the Latin American experience. The review focuses on the ways in which Latin American governments have sought to give coherence to their objectives, policies and actions and have attempted to overcome the problems arising from the need to create a wide range of specialized institutions. The final part summarizes the organizational and institutional development of government research and extension and outlines the main factors to be considered in the design of new structures which promote effective collaboration between the two activities.

What emerges from the review is a pattern of progressive institutionalization of research and extension and the links between them. The basic hypothesis of this paper is that:

- public administration becomes more complex through the incorporation of new functions in response to the emergence of new problems and the changing role of the state in development
- this increasing complexity is accompanied by an increase in interdependence between research and extension and requires greater decentralization and a corresponding decrease in the structural differentiation between the two activities, particularly with regard to adaptive and applied research

The Analytical Framework

The Nature and Context of the Problem

The linkage problems between agricultural research and extension arise from the differences in the nature of the two activities, in their objectives, and in the knowledge and resources they mobilize to achieve these objectives. Research is concerned with increasing scientific knowledge and generating new technologies. Extension is concerned with the delivery and adoption of new technologies; it relies upon communication, education and producer participation, with the overall aim of changing behavior. Thus, the two activities are carried out by professionals with different academic orientations who are answerable to different sectors of the public and whose work is evaluated according to different criteria. The fact that they are typically carried out in separate organizational contexts is an institutionalized recognition of these differences.

Integration problems are common to all organizational settings in which there is functional specialization. In the case of research and extension, the characteristics, perception and evaluation of these problems have varied from one historical period to another. The key to understanding the specific linkage problems which now exist between the two activities lies in an analysis of their institutional development.

Although institutional development is a complex process, it is not a haphazard one. It follows a general pattern, dictated by the role of the state in terms of the objectives, formulation and implementation of state policies. The role of the state in each historical situation is itself determined by:

- how power is mobilized within the state and the broader society
- the main problems confronted by the state
- the repertoire of accepted solutions
- the available economic, political, administrative, social and environmental resources

In this paper, the phrase "policy context" is used to encompass all these factors.

The analytical framework rests on an analysis of the organizational structure of public research and extension within the prevailing policy context and of the issues which define the relationships within this structure. Figure 1 (*see overleaf*) is a diagrammatic representation of this framework.

The Role of the State

The agricultural policies adopted by the state comprise the immediate policy context in which research and extension operate. These policies change from one period to another, according to the prevailing political and socio-economic objectives of the government. These objectives determine the role of the state in agricultural development, the degree of autonomy it allows agricultural institutions, the importance it attaches to research and extension and what it considers the research and extension priorities should be.

Policy objectives. The objectives of Latin American agricultural policies, and the sequence in which the policies have been introduced over time, are:

1. To develop the infrastructure needed to facilitate the production of a few commodities for export
2. To protect the producers of these commodities against production risks and market fluctuations in price and demand by imposing regulatory controls
3. To increase production and productivity, particularly through the application of new technologies, in order to reduce domestic food prices and increase export surpluses

Figure 1. The analytical framework

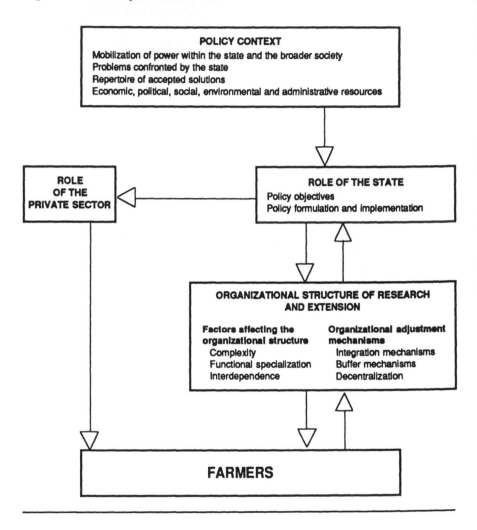

4. To reduce the possibility of social conflict by satisfying the demands
   for equity through the introduction of land reform and rural development policies
5. To develop national agricultural research capabilities, diversify production and promote social and regional development, in line with a
   new perception of agriculture's contribution to national development
   and macroeconomic adjustment policies

Policy formulation and implementation. A major factor which determines
how policies are formulated and implemented is the degree of concentra-

tion of power. To best understand the concept of concentration of power, it is useful think in terms of a continuum. At one extreme lies a total monopoly of power resources, while at the other lies equal distribution of power among all participants; the intermediate situations are those in which the degree of access to power and the opportunity to use it varies from one participant to another.

Each situation is associated with a specific decision-making process. In the first case, the subordinate levels of the government are totally responsive to the political will at the top. In the case of equal distribution of power, the overall result of the participants' decisions is independent of the will and interests of each participant. In the intermediate situations, decisions are a product of confrontation and collaboration between various groups with varying degrees of power in the decision-making process.

The rational-deductive model of decision making put forward by academics to explain the process of policy formulation and implementation is based on the assumption that there is total concentration of power. In this model, an explicit set of goals is drawn up, rational calculations are made as to how best to achieve these goals, policies are formulated and the actions required to implement these policies are defined and carried out (Dahl and Lindblom, 1953). Such a model lends support to the idea that research and extension activities can be successfully integrated through centralized control and planning. In practice, however, total concentration of power cannot exist, because of three major constraints:

- the multiplicity of participants who attempt to influence government processes and outputs
- the discretionary nature of research and extension tasks
- the limited amount of information that can be handled by those at the top of the hierarchy

Multiplicity of participants. Agricultural research and extension involves many individuals and groups of individuals, including politicians, farmers, scientists, development officers and input producers. These individuals mobilize different types of social and organizational resources to promote their interests and points of view. All government levels become the target of their pressures, and they have the capacity, in varying degrees, to block or distort public actions. This threatens the overall coherence of policies which attempt to integrate research and extension through centralized control.

Discretionary nature of research and extension. In both activities, considerable discretion must be given to the professionals at the lower levels of the hierarchy. Central planning cannot take account of all the contingencies with which these professionals are faced. The way in which they use this discretion is affected by the influence exerted upon them by different clienteles, by their personal interests and by the degree to which these interests are consistent with official policies and objectives (Etzioni, 1961).

Information constraints. The specificity, complexity and unpredictability of the problems faced by researchers and extension workers impose information constraints on the concentration of power. To integrate research and extension from the top in a centralized way would require a continuous flow of information to and from the central authorities to dictate specific commands for situations not accounted for by the central plan. In reality, however, the central authorities can handle only a limited amount of information and thus are not in a position to effectively link lower-level research and extension activities from above.

In essence, then, the total concentration of power whereby a central authority determines the activities to be carried out by the members of each subordinate unit, and has the power and ability to plan and control the integrative mechanisms which operate between these units, is a theoretical concept which rarely exists in practice. Thus, to achieve coherence, it is necessary to create linkage mechanisms which cater for the need to reach agreement on the overall objectives and to coordinate activities at the operative level.

Organizational Structure of Research and Extension

In terms of the focus of this paper, the analysis of the effect of the role of the state on the organizational structure of research and extension can be divided into: organizational factors affecting the links between research and extension; and aspects of the organizational structure which, within the confines of the prevailing policy context, can be altered to improve the effectiveness of research and extension.

Organizational factors affecting research-extension links. In the organizational structure of agricultural research and extension, the main factors which affect the links between the two activities are complexity, functional specialization and interdependence.

Complexity. The types of complexity which affect links between research and extension are:

- structural complexity (an increasing number of institutions, and hence an increasing number of mechanisms needed to coordinate their activities)
- situational complexity (a wider range of clients, commodities and environments)

- analytical complexity (an increasing number of disciplines, approaches and methodologies involved in the conceptualization of the roles and aims of the two activities)

To overcome the challenges posed by complexity, several courses of action have been put forward.

One suggestion is to establish increasingly specialized departments within research and technology transfer units, with each department focusing on a particular topic area and goal (Simon, 1976; Thompson, 1967). However, this has resulted in an increase in the interdependence *within* each unit and has not lessened the need to establish links between research and extension.

Another way of dealing with the effects of complexity is to: specify more clearly the nature and goal of each research and extension task and the way in which these tasks should complement one other; establish procedures which ensure that those carrying out the tasks conform to these specifications; and draw up rules for making decisions in potential situations unlikely to be covered by the specifications.

These courses of action incorporate a considerable degree of hierarchical control and planning. However, the greater the complexity, the less effective is hierarchical control in integrating the different contributions. Many of the difficulties which are encountered in developing effective research-technology transfer links can be explained by the fact that hierarchical mechanisms have been used. More effective ways dealing with these problems are those which rely on participative interaction and mutual adjustment at all levels (Ackoff, 1974).

Functional specialization. Functional specialization arises when, in order to complete a task, responsibilities for carrying out parts of that task are allocated to different individuals or groups of individuals.

The overall task we are concerned with here is the generation and transfer of new technologies. The knowledge, skills, functions and length of time required to develop new technologies differ from those required to deliver these technologies to the farmer.

This functional specialization forms the basis of the conventional structural differentiation between research and extension whereby the two activities are performed by different units. Inherent in such a division is a divergence in interests, values, behavior and goals between the units. In addition, the differences in the types of activities performed by each unit and in the way in which the units interact with their clients gives rise to differences in the type of management styles which are appropriate (Lawrence and Lorsch, 1967).

Interdependence. The concept of interdependence relates to the links which exist between units in order to complete the overall task. The various types of interdependence can be divided into four categories (Thompson, 1967; Van den Val et al., 1974). These categories are:

- pooled (in which the activities of one unit do not depend on the activities of other units in order to contribute to the completion of the overall task)
- sequential (in which one unit depends directly on another in order to make its contribution)
- reciprocal (in which the contribution of one unit is the input for the other, and vice versa)
- team (in which the members of one unit participate directly in the activities of the other unit, and vice versa)

There is a clear correlation between these types of interdependence and how advisors have perceived the relationship between research and extension. Initially, both activities were considered to be independent of each other but contributing to a common purpose. Later, extension was considered to be sequentially linked to research, receiving its inputs from research and incorporating these into a package of services for the farmer. Subsequently, their reciprocal interdependence was recognized, with extension identifying problems and supplying information which enabled researchers to define priorities. The current view is that the two activities should be seen as completely interdependent, with extension workers participating in experimental research work and researchers establishing closer ties with farmers.

Organizational adjustment mechanisms. The policy context in which agricultural institutions operate can rarely be altered in significant ways from the perspective of research and extension. However, three aspects of the organizational structure which can be altered to promote the effectiveness and efficiency of research and extension are: integration mechanisms, buffer mechanisms and decentralization.

Integration mechanisms. Interdependence between research and extension assumes interaction between many participants. This interaction is usually hampered by institutional mandates and restrictions, which reduce flexibility and encourage the tendency towards institutional isolation and self-sufficiency. Thus it is necessary to provide mechanisms to integrate the contributions and interests of the different institutions and the units within them.

In some cases, these mechanisms are defined and controlled by the higher levels of a hierarchy; in others, they emerge as a result of coordination between all levels of a hierarchy. The choice of integration mechanism depends on the nature of the activities involved and the degree of interdependence between them. There is empirical evidence to suggest that where these activities are characterized by a high degree of heterogenity and uncertainty, as is the case with research and extension, integration mechanisms based on coordination rather than control will be more effective (Perry, 1967; Thompson, 1967).

Buffer mechanisms. The creation of buffer mechanisms to isolate some activities from the turbulence of the environment is a defensive strategy adopted by institutions in order to reduce uncertainty and to enable them to pursue long-range goals (Thompson, 1967). Structural arrangements creating buffer mechanisms are common in the field of research and extension. Most relate to research activities, and aim to protect the long-range orientation of these activities from bottom-up demands and pressures. Others are aimed at reducing the vulnerability of both research and extension to changes in the objectives and implementation of state policies.

Extension acts as a buffer to research in that, typically, extension workers interact with farmers, passing on to them the knowledge generated by research and thus precluding a direct relationship between researchers and farmers. This is a way of "protecting" research from the immediate and diverse demands of farmers. Research can thus be carried out on the basis of a time horizon which looks beyond the farmers' immediate needs.

There are a number of ways in which agricultural institutions attempt to reduce their vulnerability to changes in policy. Among these are the constant search for institutional autonomy and the attempts to introduce automatic financing measures that would reduce the need for endless bargaining during the process of central budget formation.

Despite their advantages, the creation of buffer mechanisms is also a source of problems. The effectiveness of extension as a shield for research may lead to the latter developing along lines unrelated to the needs of farmers, or to extension itself ceasing to act as a channel for the transfer of knowledge and becoming a service for the satisfaction of other kinds of farmer needs. Institutional autonomy may undermine the capacity of policy makers to effectively orient institutional activities.

One way of overcoming these problems is to introduce an appropriate level of decentralization. Another way is to reduce the diversity of the environment with which an institution interacts by specializing in producing a specific type of knowledge (for example, by specializing in a particular commodity, discipline or region). Researchers and extension workers who focus on a specific target group or problem are less likely to lose track of policy objectives or farmers' needs.

It is also necessary to distinguish between the role and aims of basic and strategic research and those of applied and adaptive research. While the former should be protected to some degree from multiple and diverse demands emanating from the state and the farmer, the promotion of more effective interaction between researchers, extension workers and farmers implies that there should be few, if any, buffer mechanisms between extension and applied and adaptive research.

Decentralization. The aim of decentralization is to change the distribution of decision-making powers so that more decisions can be taken at regional and local level. As such, it is an issue which concerns the relationship between the higher levels of government and the agricultural institutions as well as the relationships within the institutions themselves.

It was stated above that because of the heterogenity and uncertainty of research and extension activities and of the policy context in which they operate, centralization is not a viable strategy. However, decentralization could make the various levels of the organizational structure more vulnerable to external pressures and thus have an adverse affect on overall coherence. Decentralization measures must therefore include the construction of mechanisms that will ensure control from above as well as facilitate regional and local integration.

An Historical Review of the Organization of Research and Extension in Latin America

Despite diverse historical situations and experiences, a common path followed by Latin American governments can be identified. Those whose organizational structures developed earliest, and are the largest and most complex, took the lead in incorporating organizational innovations and provided a model for other countries.

The pattern of increasing complexity in the organization of research and extension in Latin America does not differ substantially from the pattern in other developing countries. As a society matures, so new demands are made upon the state. The cumulative result of attempts to satisfy these demands and resolve specific problems is an organizational structure that is frequently disjointed and lacking in integration. In an effort to rationalize the structure, the state will often make use of organizational models developed in other contexts which are unsuited to local circumstances.

Different stages can be distinguished in this process. In the following review, the process has been divided into five stages. In each stage, the role of the state and the organizational structure of research and extension are described. The main aspects of these descriptions are then summarized in terms of the variables outlined in the analytical framework. The summaries are divided into: the variables which relate to the role of the state; those which relate to the organizational factors affecting research-extension links;

and the overall effect of both sets of variables on the organizational structure in terms of integration, the use of buffer mechanisms and the degree of decentralization.

The Early Period

Role of the state. The first stage lasted from the mid-19th century until about 1930. It was characterized by efforts to expand the state's control over the entire country, with the state becoming a revenue-collecting entity charged with safeguarding the social order and providing the infrastructure needed for the development of production, particularly in relation to export commodities. The institutional foundation for resource-expanding activities was established and state policies were oriented towards the expansion of the usable land by pushing out the frontiers, the incorporation of human resources through immigration, and the attraction of investments and resources from abroad.

This produced a geographically far-flung state based on ever-expanding legislation. With a rigidly stratified society and a simple political system allowing little participation, it aimed at establishing ground rules that would promote the accumulation of economic resources and power. Its policies reflected an "illuminist" concept of its role in society. This was the liberal period during which the state encouraged public education and science and attempted to make technologies in use elsewhere available to the productive system.

Organizational structure of research and extension. The conceptual distinction between research and extension was minimal, and thus there was no structural differentiation between them. What would now be called extension activities originated in different institutional spheres, including universities, departments within ministries of agriculture and institutions responsible to those ministries. The emphasis on education was reflected in the development of agricultural schools, with significant participation by European technicians (for example, Italians and Belgians in Peru, and Germans in Chile); these technicians played a role in the transfer of farm management techniques and carried out some adaptive crop research.

The ministries of agriculture were new and rather weak institutions. Their place within the government structure was not yet clearly defined and their efforts in technology transfer tended to be unsystematic. These efforts were incorporated into the activities of promotion units which were set up in close proximity to the farmers and had little connection with research. The units' field officers played (and in many countries continue to play) a diffuse role, acting as intermediaries between the farmers and administrative and political levels of government, as well as being involved in credit supervision and regulatory control and supplying inputs such as seeds and fertilizers; technology transfer was relegated to the bottom of their list of priorities.

Some agricultural services were initiated in response to the interest shown by farmers' associations in developing the agricultural sector. These associations had been established to defend farmers' interests and to promote their views, and they became a major source of support for the introduction of new technologies. For example, the National Agricultural Societies in Peru and Chile set up experimental stations; the Rural Society in Argentina was created with the objective of improving livestock and crop production; and, in the 1870s, the Colombian Agricultural Society was founded with the stated purpose of exchanging seeds, promoting improved livestock breeds, setting up agricultural schools and disseminating advanced agricultural techniques from abroad (Kaimowitz, 1988).

In some places a leading role was played by private sector companies interested in developing agricultural production. In the case of the railway network in Argentina, where the railway line usually heralded the productive occupation of the land, such companies provided channels for the transfer of knowledge through demonstration units scattered over the lands they owned (Marzocca, 1985). Private sector involvement was also evident among certain export companies; these companies set up technical services for the production units (such as large farms and ranches), which in many cases they themselves owned.

Summary

- A clearly stratified society, with a high concentration of power and a simple government system that facilitated the implementation of policies that had a strong influence on the economic growth of the country but were limited in number and in their degree of intervention in social and economic life. These policies required little systematic organization of the state and involved simple planning, management and evaluation procedures. Public activities were designed to promote development and agricultural production, primarily for export markets; this was reflected in their focus on the educational sphere and their efforts to meet the needs of farmers' associations which began to show interest in technical innovation.
- Low structural and situational complexity because of the existence of only a few institutions, acting completely independently and dealing with a limited range of clients, commodities and environments; low analytical complexity in that there were few distinctions between scientific disciplines, approaches and methodologies. There was

minimal functional specialization and no interdependence. Linkage problems were not perceived.

- Close relationships and relatively informal interactions between the institutions and higher policy levels because of the small size of the state apparatus. There was a low degree of decentralization. Buffer mechanisms were not necessary.

By the end of this period many countries had incipent institutions and linkage mechanisms that would develop in subsequent stages. But this development was not to be a total transformation for, in many cases, the pattern established in the early period persisted in spite of attempts to change it; in some cases it is still in evidence.

The Transition to Institutionalization

Role of the state. The second stage lasted from 1930 to the 1950s, and was characterized by a proliferation of government institutions and the emergence of a much more complex state apparatus. This situation arose because of the need to respond to the new problems emerging from the already exhausted growth potential of the development model based on the production of primary products for the export market.

As a result of market closures and the introduction of protectionist policies in the wake of the 1929 financial crisis, a regulatory state was taking shape haphazardly in some countries, notably the largest and most developed ones. Its intervention in social and economic life increased in response to the crisis, as it sought to restore the former participation of national production in international markets.

The state's ad hoc responses to the changed circumstances produced a cumulative growth in state functions. New functions gave rise to the creation of new institutions, some within the central administration and others with a greater degree of autonomy, as decentralized institutions. There were even a few instances of a certain degree of privatization in the public sphere, the control of the new institutions being the responsibility of their own specific clienteles. The case of Argentina, where a number of commodity regulation boards were created to supervise a wide range of functions aimed mainly at stabilizing markets and prices, is an example of this trend.

The first import substitution policies were implemented during the Second World War and were extended thereafter. This marked a fundamental change in the role of the state. It was no longer simply a framework for activities in which it took no direct part; instead, it became a producer, a marketing agent and a provider of credit through the development of a network of public enterprises and banks. This path, pioneered by a few countries, was later followed by almost all Latin American countries.

Organizational structure of research and extension. Research and extension were located at different levels in the state's organizational framework, and varied in terms of organizational arrangements, their degree of autonomy and their operational capacity. In some cases they were publicly funded; in others they received significant contributions from farmers. In Peru, for example, the experiment station La Molina, where an agricultural university was later established, originated in the efforts of the National Agrarian Society which had bought the land for the construction of laboratories and experiment fields and offered it to the School of Agriculture. Wherever commodity-specific activities were initiated in order to orientate and regulate production, usually with farmer participation, these activities formally incorporated technology transfer tasks.

Technological issues were gradually winning public recognition, and in the 1940s the ministries of agriculture began to undergo a thorough reorganization. For example, in the restructuring of the Chilean Ministry in 1940 and in the reforms in Argentina in 1944, a more prominent position was given to research. In Colombia, a specific agricultural ministry was reestablished in 1947. In Brazil, the National Centre for Agricultural Teaching and Research was set up to coordinate the activities of state and federal institutions.

During this period the process of creating institutional resources for research made gradual progress. In the early 1940s, Argentina had 12 experimental stations, Chile 13, Colombia four, Mexico 14, Uruguay one, and Peru a network of five stations and substations.

Commodity-centered bodies began to come to the fore, particularly those concerned with export commodities. Experimental stations established previously on the basis of farmers' contributions, such as the cotton and sugarcane stations in Peru, were now financed out of tax revenue or export duties.

Foreign assistance. Foreign technical assistance began to play a significant role. In Guatemala the National Agricultural Institute was established as a joint venture between the Chinchona Growers Association and the United States Department of Agriculture (USDA). Subsequently, the USDA initiated joint ventures in other countries, setting up experimental stations where the emphasis was on research rather than extension (Rice, 1971). By 1942, the first permanent teams of official US agricultural technicians had arrived in Latin America. Between 1943 and 1948 the USDA Office of Foreign Agricultural Relations provided technical assistance in 10 countries, with the aim of developing alternative sources of tropical agricultural supplies.

Simultaneously, US government agencies were becoming involved in the development of extension. The Institute for Interamerican Affairs (IIAA), a public entity which had been established by the US Congress to foster technical cooperation with Latin America, stimulated the creation of rural agricultural services (known as *servicios*) aimed at promoting the production of local food crops. During the war, IIAA placed teams in 11 countries. The *servicios* were bilateral, operational agencies under American direction with a network of field offices staffed by extension advisers. After 1950, the USDA programs were integrated into these *servicios* as part of a broader shift in technical assistance policies from research to extension (Rice, 1971).

Buffering research and extension. Another change taking place at this time was that those engaged in technological activities began to seek a measure of autonomy for the research and extension institutions. These institutions were responsible to the ministries of agriculture, and the intention was to protect them from the political unrest that affected the ministries. In this area, advice from the foreign agencies involved in Latin America often lay behind the moves towards autonomy, as in the case of the institutional reforms introduced in Mexico in 1945 on the suggestion of the Rockefeller Foundation.

This process was among the factors which contributed to the relative domestic isolation and highly developed external connections which began to characterize public sector agricultural institutions. Institutional relationships between countries were encouraged, again with significant participation by the Rockefeller Foundation. The Interamerican Institute for Agricultural Sciences (IICA), a branch of the multilateral Organization of American States (OAS), made an important contribution towards establishing horizontal relationships.

Originally created to offer professional training in the biological and social sciences related to rural development, the IICA played a fundamental role in introducing American organizational models and developing new approaches, particularly in relation to extension. This role was reinforced by the heavy flow of personnel who were sent to the USA for training; previously, training had been based on criteria developed in Europe and transmitted by European technicians who taught in the universities and schools.

As in the early period, there was still little or no interdependence between research and extension. The general belief was that technology was already available and that the most serious constraint on rural development was the absence of rural institutions oriented toward motivating, organizing and informing the farmers (Rice, 1971). Research units usually had their own demonstration fields and their own connections with seed suppliers. Extension concentrated on spreading cultural practices that made scant use of inputs from local research. This conceptualization meant that linkage still did not appear to constitute a problem.

Summary

- Attempts by the state to incorporate new functions in response to demands born out of new conditions in international markets; a growing number of individuals, associations and agencies interested in the development of agricultural production, with a progressive divergence of interests among them. Agricultural ministries continued to be weak in a situation in which sectoral interests were increasing their capacity to exert pressure on the state. However, there was still no real challenge to the dominant interests and perspectives which shaped policy formulation and implementation.
- Organizational structure of research and extension, and the circumstances in which they operated, becoming more complex. This was because of greater state participation in research and extension activities, the increasing influence of bilateral and multilateral aid programs and the expansion in the number and variety of institutions. In response to this growing complexity, hierarchical structures developed and there was increasing functional specialization among existing institutions. Analytical complexity increased as research and extension began to develop as different disciplines with different methodologies. There was little interdependence; institutions developed distinct identities and particular sources of financial support.
- Still a low degree of decentralization as research and extension continued to be closely connected to the hierarchical apex. However, some attempts were being made to create buffer mechanisms between the state and research and extension activities.

Developmentalism

Role of the state. From the late 1950s to the early 1970s, developmental attitudes and ideologies prevailed. In some countries, with the emergence of large urban populations, there had to be a dramatic increase in agricultural production to meet their demands for cheap food; in others, agricultural production had been declining, resulting in serious trade balance problems.

The overall policy objectives were to reform the agricultural sector and develop the rural areas. There was a dramatic increase in structural complexity: new functions were added to the established ones and new institutions appeared, many of them becoming semi-autonomous. The state was seen as the spearhead of development; it sought to determine the structure of the national economy and to redistribute power and wealth.

During this period three practices were having a strong impact on government organization: national planning, the creation of organizational systems and the introduction of development programs. The conceptualizations underlying the implementation of these practices were based on the rational-deductive model of decision making.

National planning. Planning mechanisms based on a common model became institutionalized. The influence of international bodies was evident. The United Nations Economic Commission for Latin America (ECLA) provided the technical expertise and the justification for restructuring the state apparatus, developing new concepts of the obstacles and strategies for the growth of Latin America economies. The new era of public international financing encouraged the development of criteria for project analysis and evaluation, and multilateral organizations gave advice and support in policy implementation.

Institutionalized planning aimed at being comprehensive and compulsory for the public sector. There was a conceptual and organizational separation between policy formulation and policy implementation. National and sectoral objectives would be defined by the highest political level, and would determine the criteria on which the allocation of functions were to be based. It was considered that the most effective way to achieve these objectives would be to have a clear differentiation of units, with each unit specializing in a particular type of activity. To ensure that these units functioned efficiently, a technocratic approach was adopted whereby procedures and problems were to be handled in a clearly defined rational and professional manner.

The diversity of the new functions incorporated by the government raised the problem of their coherence and integration. To address this problem, administrative control mechanisms and a series of planning levels — national, sectoral, regional and institutional — were established to ensure that planning forecasts were met. The plan provided the instruments for regulating the different kinds of interdependence, for achieving consistency between decisions at different levels and for defining the actions of the implementing agencies.

Organizational systems. With the diversity of activities for which the state was now responsible and the progressively more dynamic policy context in which it operated, new organizational structures had to be devised. The design of these structures was based on the systems approach. According to this approach, each unit is part of a whole and has hierarchical and functional relationships with other units, all oriented by the same set of goals. The design of the system consists of the identification of the appropriate structural differentiation necessary to accomplish these goals, the provision of integration mechanisms to cope with interdependencies and the establishment of a central authority with enough power and information to govern the direction of the system and the relationships within it.

In the new situation, integration could not be controlled exclusively by the top levels of the system. Each unit was given enough decision-making power to deal with the specific problems faced in the context of its mandate. Functional relationships multiplied and overall effectiveness depended on the way the various kinds of interdependence were structured. The establishment of systems as groupings of units linked by a common set of objectives was intended to solve this problem and thus facilitate policy implementation.

In the process of this restructuring, it became clear that it was no longer feasible to simply create new units within the system to meet new demands. Instead, the entire system had to be reorganized, and thus research and extension became the subject of institutional experimentation. In many cases, this radical change of thought was associated with new policies aimed at reforming the agricultural sector. An added incentive for restructuring was the availability of promising new technologies in the international arena.

Development programs. It was during this period that "Development Programs" first became a feature of national development strategies. Usually financed from abroad, they concentrated resources on the solution of problems of national priority. They were incorporated into the national plan but, to reduce the interdependencies that might detract from their effectiveness, they had a privileged administrative status and were given a certain degree of autonomy.

The fact that it was considered necessary to set these programs apart from other government activities is indicative of a growing problem: the difficulties of implementing public policies in situations of high structural complexity. Removing an activity from the central administration of the government, and internalizing within it all the functions which were necessary to achieve its goals, is a strategy now often applied to such programs in order to increase their effectiveness. The Agricultural Plan which has been operating in Uruguay since 1961, and is designed to provide technical assistance and credit, is a case in point.

Organizational structure of research and extension. From the perspective of research and extension, the most important developments during this period were: the withdrawal of US technical assistance; the creation of the decentralized semi-autonomous research institutions; and the effect of agrarian reform and rural development policies.

US technical assistance. At the start of this period, bilateral aid was still exerting a strong influence. *Servicios* had been set up through aid programs in all countries apart from Argentina, Mexico, Uruguay and Venezuela. In the smaller countries, *servicios* had started operating even before research activities developed. As late as 1966 there were 11 such arrangements, most of which played a significant role in the institutionalization of extension activities.

Within the *servicios* a process of trial and error, combined with diverging opinions within the responsible US government agencies, led to the development of a variety of organizational models. At the time, these models were the subject of a heated debate in the USA; the major issues were the relative importance and proper roles of research and extension in development aid and the potential contributions of different models to the overall process of institution building. While "productivists" tended to emphasize particular technological initiatives and productive investment projects, a group more orientated towards community development stressed the need for broader educational efforts and social and cultural change (Barksy, 1988). These debates contributed to the growing analytical complexity in the conceptualization of extension.

Withdrawal from the *servicios* began in the late 1950s, culminating years later in the end of the direct involvement of US personnel in extension activities. In their place, financial support began to be provided for specific, nationally managed extension projects (Rice, 1971).

Decentralized research institutions. The first attempts at an overall reorganization of research and extension activities were made in the 1950s. Fresh consideration was given to agriculture in a new political context, which involved governments whose banner was development, institutionalized planning and the consolidation of international public financial institutions. A new perception of problems, a developmental ideology and an assumed availability of technologies led to new approaches to research and extension.

Many attempts were made to bring coherence and order to the great variety of existing institutions. Increasing structural complexity and differentiation led to a growing awareness that many problems derived from weak or non-existent links between research and extension. The two activities began to be seen as sequentially interdependent, with extension responsible for transferring the results of research activities to the farmer.

In those countries where extension services had been established with US support, these services were incorporated into the central administrative structure. In other countries, decentralized, semi-autonomous institutions began to be established, an arrangement that eventually was to prevail in almost all Latin American countries (Piñeiro, 1983). The first of these institutions, the Instituto Nacional de Tecnologia Agropecuaria (INTA) in Argentina, was created in 1956 and incorporated both research and extension activities.

These institutions had a clear mandate. National planning was charged with issuing the policy directives that were to guide their activities and, in conjunction with joint programming, with ensuring integration between research and extension, whether they were under the same institutional roof or not. A semi-autonomous status was granted to the institutions in order to facilitate policy implementation, provide better services to farmers and develop scientific and technological capabilities, particularly with regard to advancing from adaptive research to applied and basic research.

Agrarian reform and rural development policies. The attempts at agrarian reform that took place in the 1960s placed extension activities in a new context. A new brand of institutions was born; these institutions were experimental in nature and thus their functions were not very clearly specified. Some of them were to provide a range of rural development services, including supervised credit, technical assistance and organizational support. In a few cases extension was incorporated as an additional function. This helped the development of a new conceptualization of extension activities; they were given a more dynamic orientation and focused more on social and organizational factors.

Within this context, extension was no longer sequentially related to research. A reciprocal interdependence was evident, with extension providing research with important inputs for identifying problems and defining priorities. At the same time, the new functions assigned to extension made it more interdependent with other development activities, thus multiplying the sources of influence on the extension worker and further diluting the purely technological content of his work. The difficulties that arose in the efforts to integrate different activities led to an incipient awareness that coordination through planning was inadequate. The concern to set up mechanisms to facilitate communication between the activities was growing, and this, in turn, gave rise to the new initiatives that emerged during the following period.

After 1970 a new type of policy aimed at alleviating rural poverty and increasing agricultural production was adopted by several countries. Programs of integrated rural development were implemented, most of them financed by the World Bank. These programs made heavy demands on the administrative and coordinating capabilities of public institutions, and in most cases this resulted in a duplication of functions and an increase in structural complexity.

The organizational arrangements for implementing agrarian reforms and integrated rural development programs tended to be highly centralized (Barsky, 1988). Centralization, coupled with increasing situational complexity and the multiplicity of policy instruments employed, resulted in the arrangements also becoming more bureaucratized. This reinforced the isolation of these programs from research institutions.

Summary

- A more turbulent social and political environment, with new groups claiming a larger proportion of the national income and wealth. With the development of national planning and organizational systems to cater for economic as well as scientific and technological policies, a more intricate government system began to emerge. The role of the state was not only to encourage growth but also to promote social change through transformations in the distribution of resources. This was the outcome of new demands on the state, most of them resulting from the equity problems generated by imbalances in the development pattern.
- Explosive growth of situational and structural complexity. New policies were aimed at areas and types of farmers previously excluded from public policies. A host of new institutions developed in a rather haphazard way, in spite of planning efforts. Extension activities were being carried out from a variety of different locations within the state system and relating simultaneously to various aspects of farmer problems, agricultural education and technology transfer. The intense debate on the role and aims of extension and the development of many organizational models illustrate a growing analytical complexity. Reciprocal interdependence between research and extension was apparant in many cases but, despite this, there was still effective autonomy between public institutions.
- The use of extension as a buffer, in some cases for research, and in others for central authorities in a context of growing social conflicts. Semi-autonomous institutes buffered research (and, to a lesser extent, extension) from central authorities. Integration mechanisms were revised and new forms of centralized control implemented, with growing inefficiencies in the use of resources.

The Crisis

Role of the state. In the early 1970s, most countries had relatively high economic growth rates. Strategies of development which were based on industrialization were pursued with renewed energy. Macroeconomic policies gave protection to and made possible an expansion in manufactured consumer goods, as well as intermediate and capital goods. Subsidies, rates of exchange, and trade, fiscal and food price policies discriminated against agriculture, but there were certain policy instruments such as credit and public investment that had a positive impact on agiculture (Lopez Cordoves, 1987). Government expenditure was relatively stable.

By the mid-1970s, however, the trend had begun to change. Social unrest in some countries, an energy crisis in others, and changing conditions in foreign markets for most, precipitated a crisis that was aggravated later by the impact of heavy foreign indebtedness. Since 1984, annual economic growth rates have been consistently low. All economic indicators show similar patterns: production, capital formation, employment and real wages are deteriorating, while, in most countries, inflation and trade imbalances are increasing.

In organizational terms, this crisis situation led to renewed concern for the role of planning and a change in the perception of coordination problems.

The role of planning. The greater uncertainty and instability associated with the crisis raised new problems. Without giving up its role of promoter and agent of development, the state became more like a confederation of public institutions, each with different analytical and operational capabilities and all subject to pressures from corporate interests. The increase in the diversity and scope of state objectives, and the growing inability of central authorities to control the implementation of state policies, led to a situation in which the state began to abandon the aspiration to plan everything.

It was realized that too much had been expected of planning as a mechanism for solving coordination problems. The enormous complexity arising from the proliferation of new institutions and the dynamism of the social context showed in a dramatic way the political and technical limitations of the existing planning systems. Social conflict and the uncertainties and vulnerabilities arising from the way the countries were placed in world markets challenged the assumptions of the planning techniques in use. A new, more strategic concept of the planning process started to develop.

The coordination problem. The questioning of whether the different facets of government shared a unity of purpose and consistency of action led to a change in the perception of coordination problems. Faced with decreasing resources, overwhelming short-term pressures and the weakness of the central decision-making authorities, government institutions made use of discretionary powers to reinterpret their mandates.

In the new approach to state organization, horizontal relationships were viewed as critical. The state began to be seen as similar to a market where different power groups confront each other and resources are allocated. From this perspective, it had little resemblance to the centralized, bureaucratic structure envisaged by conventional planning. Consequently, identifying and evaluating the interdependence between units became crucial to the understanding of linkage problems.

Organizational structure of research and extension. Overwhelmed by mounting pressures and problems arising from declining state resources, fractured power and external demands, the development of research and extension slowed down until, in most cases, stagnation set in. Indeed, in some countries, there had already been stagnation for some years (Barsky and Piñeiro, 1985). In addition, there was also a growing feeling within governments and among farmers' associations that public agricultural institutions were becoming excessively autonomous.

Attempts to deal with the problems arising from the crisis focused on two issues in particular: the role of extension; and centralization. The ineffectiveness of these attempts contributed to the growth of private sector involvement in agricultural technology. A separate but equally significant factor which contributed to this growth was the state's attempt to reduce internal complexity and divest itself of the some of the responsibility for agricultural technology by handing over some research and extension functions to the private sector.

The role of extension activities. The limitations of an institutional pattern based on functional specialization but confronted with groups that required multidisciplinary and multi-institutional support became dramatically clear when agrarian policies began to be more specifically targeted and to focus more on resource-poor farmers, as in the case of rural development programs. This led to a situation in which continuous organizational modifications were being imposed on extension activities; in contrast, the organizational framework for research activities remained relatively stable.

The constant search for a better coordination within and between government institutions led, and continues to lead, to ephemeral arrangements, most of them lacking the necessary administrative and political support. Extension responsibilities are distributed among various institutions which work with different types of farmers, may be associated with agrarian reform or rural development, and can be administered at either national or local level. Problems arise because of the multiplicity both of institutions and of their orientations, and differing views emerge as to how to solve these problems. Each institution seeks to legitimize its particular approach. Debates arise as to what extension means (education or technology transfer), what its goals are (empowerment, acquisition of a new set of values, or attitudinal change), who it is for (the farmer or the rural family), and how it is to be carried out (individually, through farmer groups or through mass communication).

Centralization. The model of the decentralized research institution assumes that there is a strong state with clearly defined policies in which sectoral and technological plans are consistent with the overall development strategy. The fact that this scenario seldom existed, and that research and extension activities had a wide scope in terms of types of farmers, commodities and environments, explains why in many cases the semi-autonomous institution had become an arena for tensions and conflicts in the allocation of resources. Central planning mechanisms were unable to formulate clear guidelines for identifying research and extension priorities. This role had to be assumed by the institutions, leaving other levels of the government with scant supervisory capacity.

The institutions started to play political games, searching for potential allies by distributing resources such as the location of their stations and agencies and the types of technology they generated and transferred. The main factors determining their priorities and the location and conduct of their activities were intra-institutional interests, the interests of farmers' associations and the central budgetary units of the government. These pressures were felt at all levels of the institutions, each level being influenced by varying viewpoints and interests (Martínez Nogueira, 1988).

A common response to this situation was to impose greater centralized control from the highest government levels. This had important effects on the links between research and extension and on the degree of success that each activity had in meeting its objectives.

Compared to extension, research is easier to orientate through resource allocation; it is also determined to some extent by the researcher's disciplinary background. Extension can be controlled from the center only in terms of how it is carried out; it is in the field that the actual contents of the activity, the specific relationship with the farmer and the technologies transferred are decided. These differences also reflect the different social links of the two activities. In research, pressure is exerted at the policy-making and resource-allocation levels, whereas the pressures on extension arise locally.

The role of the private sector. The private sector, in the form of farmers' associations, non-governmental organizations (NGOs) and private companies, began to play an increasingly important role during this period.

In this context, it is important to stress that access to technology varies greatly among farmers. Those who are more sensitive to technological change, and have greater resources to incorporate it into their units of production, demand more from research than from extension. They have access to private professional services, they are members of farmers' associations that provide assistance and they can purchase technologies available in the market. The small farmers come into contact with new technology mainly through government services. These differences became particularly noticeable in the late 1970s when, with the decline of most government services, more attention began to be focused on private sources of technology. As the number of private sources increased, so too did the complexity of the organizational arrangements and linkage mechanisms.

The growing importance of the private sources of technology is yet another manifestation of a recurrent phenomenon already noted in earlier periods. When farmers with resources and specific technology needs do not find an adequate response from the state, either because their aims differ from the state's aims or because they are unable to exert pressure on research institutions, they set up their own services (Kaimowitz, 1988; Piñeiro, 1983). An early example was the creation of the Colombian Coffee Growers' Federation in 1927. In Argentina, the Consorcios Regionales de Experimentación Agropecuaria (CREA) were set up in the late 1950s by medium- and large-scale farmers who felt that their need for technical information was not being adequately met by the public sector (Martínez Nogueira, 1988).

Widespread and increasing NGO involvement in agricultural technology is evident in this period. Generally externally supported, these NGOs work with poor farmers, assisting in the development of new organizations and providing a channel for the acquisition of new technical, managerial and political skills. Some NGOs, such as Central Ecuatoriana de Servicios Agricolas (CESA) in Ecuador and Centro Para el Desarrollo Social y Económico (DESEC) in Bolivia, have been highly successful; in a few countries, including Bolivia, they have even become the most important means of transferring technical knowledge to the farmer (Barsky, 1988).

Private initiatives have proved highly effective. In the case of private institutions which are commodity-oriented, the main reasons for this success are that they have homogeneous clienteles, which enables them to have a more focused concentration of activities, and their internal structure is clearly defined, which facilitates control by the center. In the case of the privately developed farmers' consortia, such as those in Argentina and Uruguay, the factors which have contributed to their success are the members' common social background, their economic resources and their fairly high level of understanding of technical matters. The effectiveness of NGOs emanates mainly from their small scale, their flexibility and their project orientation; even when they depend heavily on support from outside sources, both public and private, their legitimacy at grass roots level is well established; governments have recognized this by starting to give them a role in the implementation of programs to alleviate rural poverty.

Summary

- A dramatic increase in the uncertainties faced by Latin American societies and their governments, with a downward trend in the levels of economic growth. This was accompanied by a decrease in the state's capacity to orient society and promote development and a tendency to hand over tasks to the private sector. Crisis management prevailed, with the state's efforts focused on pressing short-term problems. Planning systems were reorganized, with a progressive loss of confidence regarding their effectiveness and their ability to adjust to new circumstances. Resources declined, putting additional strains on public institutions; the state attempted to handle this by making frequent attempts to reduce the autonomy of implementing agencies. Expansion of public institutions was curtailed, resulting in a decrease in their scientific and technological potential.

- An increase in structural complexity, resulting from the creation of new organizational arrangements to deal with specific problems and sectors of the population and from the consolidation and expansion of private and quasi-government involvement in agricultural technology. Situational complexity continued to increase as a result of the uneven impact of the crisis on different commodities and type of farmers. The frequent modifications of the organizational framework for extension were symptomatic of growing analytical complexity. Inherent in some of these measures was a recognition of reciprocal interdependencies and hence the need to reduce functional specialization.
- Inability of extension to play its buffer role because it was the subject of frequent reorganization in attempts to determine its role and location in the organizational structure. This sometimes led to research attempting to by-pass extension altogether and link directly with farmers. The growing attempts to centralize decision making dislocated links between research and extension.

The Present

Role of the state: New challenges. The policy context which determined the role of the state during the crisis period persists. However, although there are some efforts being made to design specific organizational structures which suit this context, more attention is now being paid to the development of general ideas and proposals which will meet the challenges emerging from the crisis situation.

These challenges have a dual nature. On the one hand, public institutions must become more efficient and effective because of the pressing situation in terms of the resource base and the need for less costly and more productive technologies. On the other hand, while many new private institutions have been established and have provided alternative means of generating and transferring technology, their actions must be made consistent with national priorities, and efforts must be made to ascertain how their activities can complement those of the public institutions.

The most critical issues in the current situation are:

- the exhaustion of the development model which has oriented most state policies since the Second World War
- the serious problems arising from the foreign debt situation, which

necessitate a renewed effort to diminish the burden of agricultural imports and to heavily promote exports
- the state's financial crisis, which requires a reorientation of public spending
- the introduction of adjustment policies in an attempt to overcome current difficulties and reorientate the economies

In the face of these problems, government policy is best oriented on the basis of strategic proposals rather than on the basis of a set of detailed, comprehensive plans; the latter course is being abandoned almost throughout the region. There is a heated debate as to the role and functions of the state, and attempts are being made to focus action, make better use of scarce resources, improve performance and redefine the role of the private sector.

Impact of new challenges on the organizational structure of research and extension. Agriculture has taken on renewed importance, and is now being seen as a strategic means to alleviate current difficulties. Against this background, various trends can be identified:

- increasing concern about the technological lag that could occur in Latin American countries in relation to research progress being made elsewhere, and about the private appropriation of the results of this research
- consolidation and expansion of private efforts in the field of technology generation and transfer
- questioning the appropriateness of the prevailing institutional model, with its centralized nature and lack of flexibility

In response to these trends, the national systems have undergone various kinds of reorganization in an attempt to increase decentralization, improve efficiency in the use of resources, acquire new sources of finance, introduce greater flexibility and derive maximum benefit from accumulated scientific capabilities. Major organizational changes are currently under way in Argentina, Colombia, Ecuador, Mexico and Uruguay.

The debate on extension has continued, to the point where even the term itself has fallen into disuse. Greater emphasis is now being given to what is specifically connected with technology transfer, and it is in this area that most new institutional arrangements are occurring. In Chile, for example, the Instituto Nacional de Investigacion Agropecuaria (INIA), has followed the example of the CREA groups in Argentina and established its own technology transfer groups (GTTs). Comprised of farmers, these groups exchange their experiences of different technologies with each other and receive technical support from INIA; once they reach maturity and can operate with minimal support, INIA partially disengages from them (Goldsworthy and Kaimowitz, 1988).

Efforts to increase the relevance and responsiveness of research and improve its links with extension include conducting on-farm research and redesigning the functions of the experiment stations. On a broader scale, it has become clear that certain features of the organizational structure, particularly those relating to increased centralization, are unworkable in the present situation. The desire to achieve adequate control through centralized coordination mechanisms clashes with the ability of powerful interest groups to persuade high-level policy makers to meet their demands. Overloaded by detail, the state loses sight of strategic issues and becomes enmeshed in bureaucratic procedures. This weakens its administrative and information dissemination capabilities, inhibits regional and local adaptation of policies and restricts the capacity for mutual adjustment between the institutions implementing these policies. The institutions begin to compete with each other and to develop contradictory aims and interests. In such a situation, coordination mechanisms become inoperative.

Thus, it has been necessary to change course. In the debate as to how to do this most effectively, certain major issues have come to the fore—decentralization, farmer participation, cooperation between public and private institutions, and interaction between research and extension.

Decentralization. During the crisis, a common policy was to try to reduce the vulnerability of research and extension to external pressures by reinforcing centralization within institutions. Now, new approaches are necessary to decentralize these activities while preserving an institution's ability to plan, monitor and evaluate them.

The experience of INTA in Argentina provides an example of why a new approach is needed. Originally, a central extension unit provided the general orientation of extension activities. In terms of day-to-day activities, however, the extension agents were responsible to INTA's experiment stations; these stations also managed the research programs and tended to relegate extension to second place. Their supervision of extension activities was reduced to bureaucratic procedures and the extension agent had to use his own initiative, within the broad guidelines provided by the central unit, as to how to dispose of the insufficient resources at his command. Thus, there was vulnerability in the field and heavy dependence on the center for direction.

To overcome such problems, attempts are being made to design an organizational structure that is both responsive to farmer's needs and oriented by national policies and priorities. These new structures are not based on centralized decision making, and central planning is evident only in terms of strategic issues. Planning is seen as a process in which the core features are participation by all relevant groups and emphasis upon the evaluation of research and extension activities and their impact upon the farmer. In other words, qualitative, rather than quantative, planning will determine the direction of research and extension.

Decentralization is an organizational resource for coping with structural and situational complexities. Decision-making powers are located at the lower levels of institutions and closer to the farmer. To be effective and to foster responsivesness from the farmers, administrative procedures have to be simplified. However, there is a risk that autonomous institutions which lack a common purpose will develop. To reduce this risk, decentralization must be complemented by other institutional mechanisms.

Farmer participation. If responsiveness is desired, a systematic channeling of farmers' demands through farmer participation in planning and decision making is required. In this way, farmers' needs can be incorporated into the problem identification, priority setting and progamming stages.

However, as in the case of decentralization, there are risks attached to this strategy. One is the possible corporate takeover of farmers' organizations. Another is the possibility that research and extension institutions will become arenas for resolving intrasectorial conflicts; if the interests and demands of farmers who are pressurizing the institutions are not homogenous, research will be pulled in one direction, extension in another. Thus, integration between the two activities is adversely affected. This is less likely to occur in situations where research and extension have a narrow commodity or geographical focus; the greater the disparity in the production systems with which the activities are confronted, the more likely it is that farmer participation will threaten integration.

Cooperation between the public and private sectors. Technology transfer is carried out not only by public institutions but also by private agencies, and problems arise in the attempts to set up operational coordination mechanisms between the two sectors. How to establish an effective cooperative relationship between public and private institutions (such as NGOs, private companies, cooperatives and farmers' organizations), with a clear division of roles and well-defined complementarities, is an issue that has assumed a new relevance in the light of declining state resources and growing technological awareness among farmers.

In many cases, to establish this relationship requires a change in orientation, attitudes and methods and a redesign of the prevailing linkage mechanisms. Public institutions have paid only slight attention to private sector development of new scientific and technological capabilties, partly because of the public sector's traditional bias against the private sector. The results of public research have been transferred through a limited number of channels, without the involvement of the private sector.

Public institutions must develop methods to bring researchers into closer contact with the farmer and make research results accessible to all those involved in technology transfer (Bunting, 1986). Extension services must be more flexible, in recognition of the fact that there are many different types of farmers and thus many different ways of transferring technology, and the activities of extension workers must reflect a greater awareness of the services being provided by private institutions. Whereas commercial services and collaborative arrangements are becoming more widespread among larger farmers, cooperatives and NGOs are providing services to small farmers, thus complementing the work of traditional extension services.

It is likely that the outcome of adjustments to take account of these factors will result in the establishment of *networks* of institutions, linked by a variety of mechanisms.

Interaction between research and extension. The new challenges to the organizational structure of research and extension have highlighted the need to establish more effective interaction between the two activities. In order to meet this need, revisions must be made in three particular areas: use of financial and material resources; evaluation criteria; and personnel management.

Links at the local level require a flexible use of resources, but public sector regulations limit this flexibility by inhibiting the exercise of discretion at this level. These constraints also limit participation in decisions concerning resource management. This rigidity must be overcome.

The differences in the nature of research and extension means that follow-up and evaluation are usually based on different indicators. Whereas research is evaluated on the basis of conventional scientific criteria, the evaluation of extension is concerned with the effective transfer of technology. These differences give rise to different routines and behaviors aimed at satisfying the evaluation criteria. New evaluation criteria, which combine relevance, adequacy and transferability, must be developed.

Public sector personnel management regulations require a clear definition of the powers and obligations of each staff member. Currently, the definition of a researcher's role is much less controversial than the definition of the role of the extension worker. Associated with these definitions are important salary, promotion, and career development differences between researchers and extension workers, which contribute to the lack of effective interaction. Policies related to these issues must be revised.

Conclusions

A general pattern of progressive institutionalization of research and extension emerges from the historical review of public sector agricultural research and extension in Latin America. From this pattern, certain conclusions can be drawn as to the main factors which should be taken into account in the design of new organizational and institutional structures.

Organizational and Institutional Pattern of Development

As indicated in the introduction to the historical review, the Latin America experience does not differ substantially from that in other developing regions. However, it must be borne in mind that in some respects the heritage of government organizations and the challenges that they face differ from one region to another, and thus some elements of the pattern outlined here are peculiar to Latin America.

The pattern of organizational and institutional development in relation to agricultural research and extension is as follows:

- The initial situation is characterized by the existence of a variety of independent institutions, each institution fully exercising its own decision-making powers and pursuing its own objectives; there is little distinction between research and extension. There is minimal complexity, interactions are random and linkage does not seem to constitute a problem.
- The number of institutions participating in research and extension activities increases, and there is a corresponding increase in structural complexity and functional specialization. The state attempts to reorganize the structure by placing both activities under the same authority. Research and extension are seen as forming part of a functionally related whole; this pooled interdependence is controlled through hierarchical arrangements aimed at achieving coherence.
- The objectives of the state policies expand. As a result, more institutions, with a wider range of client groups, commodities and environments, are created and there is greater differentiation between research and extension. This increase in structural and situational complexity leads to a change in the concept and objectives of extension; it is first seen to be sequentially related to research, but later a reciprocal relationship is evident. National planning attempts to achieve compatibility between the policies governing the two activities and between these activities and sectoral objectives. The organizational structure is redesigned, based on the systems approach, in an attempt to orientate institutions towards a common goal and coordinate their actions through centralized control.
- Complexity continues to increase, and the inadequacy of previous arrangements becomes evident. The nature of research and extension and the diversity of the environmental conditions they confront

renders hierarchical arrangements ineffective in terms of orientating the activities. The attempt to monitor activities leads an overflow of data at the upper levels of the organizational structure. Centralized planning mechanisms fail to promote effective linkage at operational levels. In an effort to provide specific responses to the variety of economic and social demands, there are instances of some links being created at the operational level, based on mutual adjustment between activities.

- In the context of increasing complexity, the state recognizes the need to improve the efficiency and effectiveness of research and extension, to introduce a greater degree of flexibility and to derive maximum benefit from accumulated scientific knowledge. This, in turn, leads to a recognition of the need for constant interaction between research and extension, and hence less differentiation between the two activities. A major reorganization becomes necessary. The central features of this reorganization are an increase in decentralization, farmer participation, cooperation between the public and private sectors, and the establishment of interactive linkages between research and extension. Associated with this process, new modes of action are developed, including on-farm research, researchers working with technology transfer groups, and the allocation of some research responsibilities to extension workers.

Design of New Organizational and Institutional Structures

The historical review shows that the design of new organizational structures should take into consideration the policy context in which they operate and the type, source and frequency of the interactions research and extension institutions are expected to develop with the different sectors of society. Contingency theory, a well-developed perspective in organizational analysis, postulates that an institution's effectiveness depends on the "goodness of fit" between policy context and organizational structure.

This implies that prior to reorganization, the policy context should be analyzed. The more fragmented the power structure and the larger the number of people involved in formulating and implementing research and extension policies, the greater the complexity of the organizational structure will have to be. Given this greater complexity, the most effective reorganization should incorporate a reduction of differentiation between research and extension at lower institutional levels which are closer to the farmers while, at the same time, creating buffer mechanisms to protect basic and strategic research activities from short-term demands and political uncertainties.

The evaluation and selection of organizational structures is influenced by the conceptualization of the state's ability to formulate and implement coherent policies. Thus, to design appropriate structures, an analysis must first be made of the assumptions underlying these conceptualizations and of how these assumptions conflict with reality; the analytical tools used must be capable of capturing the complexity of the environment and of the interaction of research and extension with this environment. In general, a modest view of the state's managerial and administrative capabilities should be adopted, with planning being guided by strategic considerations at the central level and fostering participation and interaction at the operational levels (Crozier, 1987). To establish more effective interaction, the design of new structures must take into account the need for greater flexibility in the use of resources, for coherence in the evaluation of research and extension activities and for personnel management policies which incorporate a clear definition of roles and more equitable terms and conditions of employment between researchers and extension workers.

In terms of the design of the institutions themselves, the choice consists of various ways of allocating tasks, of regulating interdependencies and of allowing for complexity. It has been shown that excessive differentiation of research and extension, based on functional specialization, has resulted in severe linkage problems. However, the nature of the two activities makes a certain degree of differentiation unavoidable. Thus, institutional design should aim at achieving an adequate balance: differentiation should incorporate the recognition of the particular knowledge and skills required in the generation and delivery of technology, but it should not be so great as to isolate one part of the process from the other. It is important to remember that research is not a homogeneous activity. While, on the one hand, a logical step towards avoiding linkage problems may be to reduce the differentiation between extension and adaptive and applied research, on the other a differentiation between these activities and basic and strategic research may be necessary to protect the latter from everyday demands.

In the light of a growing understanding of the nature of the interdependence between research and extension, the design of linkage mechanisms and the perception of linkage problems has changed. Team interdependence, whereby the technological needs and priorities in a given situation are established as a result of close collaboration between research and extension, is now seen as the most effective way of carrying out these activities. Thus, institutional design should allow for mutual adjustment at operational levels, and this implies that the decentralization of certain decision-making powers must be undertaken. The need for decentralization is reinforced by the high levels of complexity which now characterize the organizational structure of research and extension and the policy context in which they operate.

Acknowledgments

I owe a considerable debt to David Kaimowitz, from ISNAR, for his comments and advice on the content and form of this paper. I would also like to thank Marie-Helen Collion, Huntington Hobbs, Deborah Merrill-Sands and Willem Stoop, all of whom work at ISNAR, and Kay Sayce, of Chayce Publication Services, for their assistance in drafting this paper.

References

Ackoff, R. *Redesigning the Future.* New York: John Wiley, 1974.

Barsky, O. *Las Políticas de Desarrollo Rural en América Latina: Balance y Perspectivas Estratégicas.* Mimeo. Buenos Aires: CISEA, 1988.

Barsky, O., and Piñeiro, M. *Evolución de la Productividad y el Cambio Técnico en el Sector Agropecuario de América Latina.* Mimeo. Buenos Aires: CISEA, 1985.

Bunting, A. H. "Extension and technical change in agriculture." *Investing in Rural Extension: Srategies and Goals.* Jones, G. E. J. (ed). London: Elsevier, 1986.

Crozier, M. *Etat Modeste, Etat Moderne.* Paris: Fayard, 1987.

Dahl, R., and Lindblom, C. *Politics, Economics and Welfare.* New York: Harper and Row, 1953.

Etzioni, A. *Modern Organizations.* Englewood Cliffs: Prentice Hall, 1961.

Goldsworthy, P., and Kaimowitz, D. "Las relaciones entre la investigacíon agropecuaria y la transferencia de tecnología: El caso de Chile." The Hague: ISNAR, 1988.

Kaimowitz, D. "Agricultural technology institutions in Colombia and the linkages between research and technology transfer within them: An introductory overview." ISNAR Staff Notes. The Hague: ISNAR, 1988.

Lawrence, P., and Lorsch, J. *Organization and Environment.* Boston: Harvard University Press, 1967.

Lopez Cordoves, L. "Crisis, políticas de ajuste y agricultura." Sgo. de Chile *Revista de la CEPAL* (33, 1987).

Martínez Nogueira, R. "El apoyo político a la investigacíon agropecuaria en América Latina." *Temas Prioritarios y Mecanismos de Cooperación Agropecuaria en América Latina y el Caribe.* Cali: CIAT, 1988.

Marzocca, A. "Proceso de formación y evolución del INTA en Argentina." *Organización y Administración de la Generación y Transferencia de Tecnología Agropecuaria.* Stago, H., and Allegri, M. Montevideo: IICA, 1985.

Perry, C. "A framework for the comparative analysis of organizations." *American Sociological Review* (32, 1967).

Piñeiro, M. *Articulación Social y Cambio Técnico.* San José: IICA, 1983.

Rice, E.B. *Extension in the Andes.* Cambridge, Massachusetts: MIT Press, 1971.

Simon, H. *Administrative Behaviour.* New York: Free Press, 1976.

Thompson, J. *Organizations in Action.* New York: McGraw, 1967.

van den Val, A., and Delbeck, A. "A task contingent model of work-unit structure." *Administrative Science Quarterly* (19, 1976).

4

Intergroup Relationships in Institutional Agricultural Technology Systems

Paul Bennell

Links between agricultural research and technology transfer can be analyzed from a number of perspectives, including economic, political, organizational and socio-psychological perspectives. To date, much of the discussion on the links in developing countries has focused on organizational issues and much less on the human behavior that shapes the relationships between agricultural research and technology transfer personnel. However, as Frosch points out, "the single most important point about technology transfer is that it is a human process.... Unfortunately, it is frequently treated as an organizational difficulty, or a communications problem, to be dealt with by the establishment of the appropriate reports, planning systems and organizational modes" (Frosch, 1984). This observation is based on the findings of a survey of relationships between research and development (R & D) groups in industrial corporations in the USA, but it appears equally valid for agricultural research and technology transfer groups in developing countries.

The purpose of this paper is to investigate the potential impact of various behavioral and socio-psychological processes on the relationships between agricultural research and technology transfer personnel in developing countries. Building effective links within institutional agricultural technology systems (IATS) requires that the individuals participating in these systems have certain values, attitudes, beliefs and goals which motivate them to relate to each other as well as possible. Normally, each IATS comprises two major groups of individuals (typically "researchers" and "extensionists") who interact in various ways. From a socio-psychological perspective, my primary concern is, therefore, with the perceptions, motivations, feelings and actions of these individuals, in order to identify how they influence, and are affected by, the relationships between the two groups. Social psychology — the study of behavior in social contexts — is able to offer considerable insights into the nature of these relationships.

The paper is structured as follows. The first section, "Links between IATS Groups", briefly discusses the types of relationship which are officially prescribed, which are theoretically possible, and which actually exist — they are usually different — between the two groups. A simple model of the organizational structure and associated occupational subgroups commonly found in public sector IATS is presented in order to facilitate the analysis. The second section, "Relationships between Groups", considers the distinction between interindividual and intergroup relationships. In the third and fourth sections, I examine the usefulness of the two main socio-psychological theoretical approaches to intergroup relationships, Realistic Conflict Theory and Social Identity Theory, for explaining relationships between groups within IATS.

Other relevant characteristics of groups not specifically covered by these theories are dealt with in the fifth section, "Intragroup Characteristics and Intergroup Contact". The sixth section, "Occupational Groups and Structure", focuses on more strictly sociological explanations of occupational groups and their inter-relationships. In final section, I summarize and draw some conclusions.

Links between IATS Groups: An Overview

The Desired Relationship

The importance of establishing effective research-technology transfer relationships has been constantly reiterated by developing-country politicians, policy makers, IATS leaders and other experts in official statements and policy documents, as well as at national and international seminars and other meetings specially convened to discuss links within IATS (for example, *see* FAO, 1979; Asian Productivity Organization, 1980; Cernea et al, 1984).

The linkage requirements of IATS are determined by the specific nature of each technology generation and technology transfer subsystem. In general terms, however, individual members of these subsystems must interact with each other to perform all or some of the following functions:

- diagnosis of producers' problems
- testing of new technologies
- communication of new information from research to technology transfer

- feedback of information about the performance of agricultural technologies from technology transfer to research
- training, mainly of technology transfer staff by researchers
- provision of services and regulatory functions (for example, seed testing, pesticide registration, soil testing, plant pathology services)

In developing countries, it is the third function, the communication of information from research to technology transfer, that has received most attention from public sector institutions. However, increasing emphasis is now being given to the need for more intensive collaboration between researchers and technology transfer workers in developing technology packages for producers. Lionberger refers to this process as the "information integration" function (Lionberger and Chang, 1970; Lionberger, 1986). Consequently, the testing and feedback functions have assumed greater importance.

This has critical implications for the nature of links within IATS. In the past, links were characterized by sequential, functional interdependence, in that the research group was seen primarily as providing knowledge and technology to the extension group which, in turn, transferred these to farmers. However, in operational terms neither group impinged significantly on the other. The new emphasis on information integration means that links within IATS are now conceived in terms of reciprocal, task interdependence. That is, neither research nor extension can fulfill its responsibilities without the operational involvement of the other.

Given the high degree of interdependence between the two groups it is usually argued that links should be strongly collaborative, facilitated by good communications, and with the members of each group well motivated and endowed with the necessary skills to interact in the prescribed fashion. Feelings and attitudes of trust, mutual respect, empathy and understanding should therefore underpin linkage relationships.

Possible Relationships

The types of relationship between IATS groups that are theoretically possible can be categorized as follows:

- cooperative-collaborative, as is normally officially prescribed by IATS, where both parties are strongly motivated to interact effectively with one another
- competitive-collaborative, where there is considerable competition between the groups but where each group's need to collaborate with the other in key areas enables workable compromises to be reached on

the basis of constructive bargaining (cooperative- and competitive-collaborative relationships should be seen as part of a continuum rather than as pure categories)
- conflictual-engagement, associated with negative feelings and attitudes (such as hatred, dislike, mistrust, annoyance, condescension, disdain, jealousy, lack of respect) and where group members openly engage in conflict with little scope for compromise
- conflictual-avoidance, again associated with negative feelings and attitudes, but where one or, more typically, both groups avoid interacting with each other to varying degrees
- indifference on the part of one or both groups; here interdependence between the groups is perceived as zero or insignificant

This is a useful categorization of possible relationships between groups in IATS. In practice, however, relationships are often more complicated than this. For example, the categorization implies that each group has symmetrical views about the other. While collaborative relationships are generally based on symmetrical motivational needs of both groups, relationships in which the motivation for one group to interact with the other is asymmetrical can also exist. In this situation, one group is dependent on the other and is therefore motivated to collaborate, whereas the other group has little or no motivation and is largely indifferent. The nature of the relationship depends on the distribution of power between the two groups but, given that one group is unwilling to collaborate, the outcome is likely to be conflictual.

It is also possible for different combinations of relationships to co-exist with respect to different functions or individuals: "Two organizations may engage simultaneously in different types of cooperation or contest interaction focused on different issues and involving different configurations of personnel, interest, domain sectors, and types of resources applied to the issues" (Thomas et al., 1972).

The Relationship in Practice

A review of the recent literature on links in IATS, including papers given by IATS personnel at international and national meetings, shows that the research-extension relationships in developing countries rarely correspond to expectations or officially prescribed norms, given the nature of the alleged interdependence between the two. That is, links are only exceptionally cooperative-collaborative or even competitive-collaborative. Palmer and his associates, for example, no doubt heavily influenced by their Central and South American experience, suggest that conflictual-engagement relationships are the norm between research and extension organizations in developing countries:

> The research-extension relationship is often poor. This situation has resulted over many years as a result of competition for funds, manpower and physical facilities and the perception of each service's performance by the other. Administrative and budget structures generally discourage rather than encourage communication, cooperation and integration of the two services. There is often little or no contact and a lack of respect between the two groups. Often there are power struggles between the heads of the two agencies (Palmer et al., 1983).

I argue, however, that such openly conflictual relationships are in fact fairly rare among IATS in developing countries. The available evidence suggests that relationships characterized by conflictual-avoidance and indifference are more common. As Compton remarks, "the biggest problem may be that such strains and tensions (between research and technology transfer groups) exist but there is a general failure to recognize or accept this fact.... The failure to see the need for more effective interactive exchange between research scientists and extension personnel is of greater concern" (Compton, 1989).

A high level of indifference is borne out by two recent surveys of research and extension staff. In a questionnaire survey of extension directors in 59 countries, Sigman and Swanson found that "contrary to the general expectations of the literature, only 17% and 16% respectively of those responding perceive linkages and technology as serious problems" (Sigman and Swanson, 1982); the directors ranked links eighth out of nine possible factors affecting their institutions' performance.

A survey of Indian research scientists, carried out by Balaguru and Rajagopalan, found that the scientists ranked an efficient extension service as the least important of 12 factors determining their research output. The authors concluded: "It is rather strange to note that the agricultural scientist gave least importance to the extension service, which is considered the hallmark of agricultural research output in the country" (Balaguru and Rajagopalan, 1986). Rather than being strange, this situation is probably quite common in developing countries. I shall return to this in the section on Realistic Conflict Theory.

Poor links are not confined to IATS in developing countries. Problems in the relationships between research and development and/or between marketing and production departments in industrial organizations in both developed and developing countries is a major theme in R & D management literature. Furthermore, despite the differences between organizational environments, many factors seen as responsible for poor links in industrial organizations are very similar to those which undermine relationships between IATS groups in developing countries.

However, there are two factors confronting these IATS which distinguish them from their counterparts in the developed world. First, they are often institutionally much weaker, with limited human and financial resources and frequently with inadequate political support. In addition, many IATS institutions are relatively young, particularly in countries which have gained political independence in the past 20 to 30 years.

Second, the status and educational and cultural gaps between researchers, extension workers and farmers are usually considerably larger in developing than in developed countries. I shall return to this issue in the sixth section of this paper.

Organizational Structure and Group Composition

IATS in developing countries have a variety of organizational structures. The most common structure within the public sector is the existence of separate research and extension departments, located in the same organiza-

Figure 1. Group composition and links between subgroups in the two-department single-apex management structure

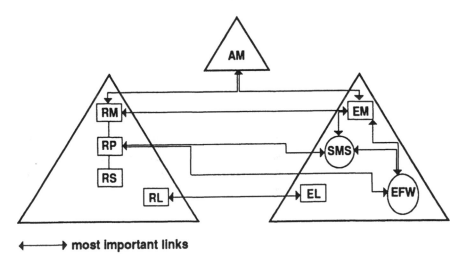

◄──────► **most important links**

AM apex management; RM research management; RP research professional; RS research support; EM extension management; SMS subject-matter specialist; EFW extension field worker; RL research liaison; EL extension liaison

tion (a ministry, parastatal or university) or the responsibility of separate organizations (Swanson and Rossi, 1981). Here, I shall focus mainly on the two-department structure with both departments controlled by a single policy making/management group, which I shall refer to as apex management (AM).

Even within this fairly simple structure, intergroup relationships are likely to be complex. In a typical research organization, three main subgroups can be identified: the research managers (RMs), the research professionals (RPs) and research support (RS), the latter including all technical and administrative staff.

Extension personnel can be divided into the following subgroups: the extension managers (EMs), the extension specialists (subject-matter specialists, or SMSs), and the extension field workers (EFWs). In addition to these, some IATS, especially the larger ones, have specific liaison positions, including liaison officers and information and training specialists. These people may constitute distinct subgroups, research liaison (RL) and extension liaison (EL).

In most discussions of the links between research and technology transfer within IATS, the greatest amount of emphasis tends to be placed (at least implicitly) on the RP-SMS relationship and the RP-EFW relationship. However, as Figure 1 shows, there are at least another five intergroup relationships which have a key bearing on the effectiveness of links between IATS groups.

These intergroup relationships are:

- the RM-EM relationship
- the separate relationships of EM and RM with AM (that is, the EM and RM relationships with government ministers, deputy ministers, permanent secretaries, directors general and their senior assistants)
- the SMS-EFW relationship (extension departments are increasingly seen as being part of the agricultural research system, to the extent that SMSs are becoming heavily involved in adaptive and systems research; while this may strengthen the integration of the IATS as a whole, it may weaken links between SMSs and EFWs; one of the most critical linkage relationships at present is thus *intra*-departmental)
- the RL-EL relationship

These relationships need not be identical and may even be highly diverse, ranging from cooperative-collaborative relationships to those which are characterized by indifference (*see* Figure 2 *overleaf*). Thus, each subgroup's relationships should be studied independently. However, the relationships between the management subgroups are likely to be critical in setting the tone for those of the other subgroups.

Figure 2. Possible relationships between IATS subgroups

Type of relation-ship Subgroups	Cooperative-collaborative	Competitive-collaborative	Conflictual-engagement	Conflictual-avoidance	Indifference
AM — EM					
AM — RM					
RM — EM					
RP — SMS					
RP — EFW					
SMS — EFW					
RL — EL					

Relationships between Groups

In analyzing the relationships between groups, a precise distinction must be made between interindividual and intergroup behavior and relationships. Some socio-psychological theories place primary emphasis on the role of interindividual relationships in explaining relationships between groups. In contrast, the dominant theoretical perspective of contemporary social psychology focuses on the existence of intergroup relationships and behavior which, because they are the outcome of group processes, are quite distinctive from those governed by strictly interindividual relationships.

Sherif defines intergroup behavior as occurring "whenever individuals belonging to one group interact, collectively or individually, with another group or its members *in terms of their group identification*" (Sherif, 1966; italics mine). Similarly, Turner argues that "the perception of ourselves and others is more influenced by group memberships (rather than by personal identity) in some contexts. The transition in cognitive functioning from personal to group identity corresponds and underpins a shift from interindividual to intergroup behavior" (Turner, 1982).

Interindividual Relationships between Groups

From this perspective, the main concern is to identify the personality characteristics and the values and attitudes of individual group members which exist independently of interaction between groups.

Individual personalities. At the simplest level, the linkage relationships within IATS can be explained in terms of the interaction of individuals. If the balance of personalities is favorable, then good relationships will prevail.

Certainly, a common perception among IATS personnel and senior managers in developing countries is that poor links are often the result of personality clashes. If this is so, then presumably removing key individuals in extension and research who cannot get along would go a long way toward overcoming poor relationships between groups.

Individual personalities are likely to be an important factor in shaping relationships between IATS groups in certain situations. For example, Akinbode, in his study of agricultural research and extension links in western Nigeria, concluded that personality factors were the most frequent cause of conflict between the two groups (Akinbode, 1974). Nor would it seem that this factor becomes any less important with increased economic development: in a survey of R & D organizations in the USA, Souder concluded that "many distrust cases (between individuals in R & D and other groups) are characterized by personality conflicts" (Souder, 1980).

It is quite common for social conflicts to be attributed to personality clashes, both by observers and by participants. However, this still begs the question whether individual personalities are the real cause of conflict, or whether they are merely a symptom of poor relationships between groups as a whole.

It is well known that people with broadly similar attitudes are more likely to get on with each other than those who hold different views. If mutual attraction is a component of a successful working relationship, then good links between IATS groups will depend on ensuring that EMs and RMs and other research and extension personnel who deal with each other have similar attitudes. However, a major weakness of this argument is that it does not address the question of why people from different groups might have different attitudes in the first place. In other words, the very disparity in attitudes which is thought to be causing the problem may well originate from some group or intergroup source.

Unfortunately, no empirical evidence has been collected on whether there are systematic differences in the attitudes of research and extension personnel that are independent of group and intergroup processes. It is conceivable that people with different views and attitudes are attracted to different occupations and hence different groups. For example, it is sometimes argued that research scientists tend to be individualistic, and are therefore loners. As such they would be less likely to establish cooperative relationships with technology transfer personnel.

Personality, culture and modernity. One of the major preoccupations of anthropology is to identify the effect of specific aspects of an individual's culture in shaping his/her personality and/or behavior. Some of the main personality characteristics that have been shown to be culturally influenced include need achievement, aggression, dominance, conformism, anxiety, self-esteem and cooperativeness. Such characteristics, if they are widespread within a population, could be important in explaining the nature of intergroup relationships within IATS.

A society's culture and structure shape social relationships and behavior. Again, this has critical implications for relationships between IATS groups. For example, the more socially stratified, authoritarian and/or status-conscious a society is, the more difficult it is likely to be for individuals from different classes, status groups, castes or ethnic groups employed by an IATS to collaborate effectively. Leonard's study of agricultural extension in Kenya showed how formal organizational relationships were frequently supplanted by behavior based on informal, political and ethnic patron-client relationships (Leonard, 1977). Cultural attitudes and behavior concerning relationships between people of different ages and sex are also potentially significant.

According to modernization theory, poor relationships between IATS groups in developing countries can be directly attributed to the extent to which individual employees have internalized "modern" values and attitudes. This is because "all forms of complex organization entail inconsistencies with prevailing norms and values in traditional societies more than in modern ones" (Tannenbaum, 1985). Modern values and attitudes are characterized by high achievement needs and a strong attachment to occupational roles (professionalism). However, modernization theory is too vague to be of much value in getting to grips with the complexities of intergroup relationships within IATS.

In the absence of hard evidence, it is difficult to assess the role of interindividual relationships in shaping linkage relationships in IATS. I hypothesize, however, that interindividual relationships are more likely to be an important contributory factor if: (1) the group identity of researchers and/or technology transfer personnel is low; (2) there are relatively few individuals in each group (as is the case in small countries); and (3) both groups work in close proximity to each other, especially in remote locations. A key indicator of the importance of interindividual relationships would be the degree of their variability, that is, the absence of stereotyping.

Intergroup Relationships

As noted earlier, the group rather than the individual is now the main unit of analysis for theories of intergroup relationships. Groups are seen as distinctive socio-psychological entities which condition their members' perceptions of the world around them, and especially their relationships with other groups. This is thought to be a collective process that is not only *not* the sum of the individual attitudes and actions of group members but may even depart considerably from some individual preferences.

IATS comprise specific occupational groups whose members, by definition, share the same or similar organizational positions, participate in equivalent work experiences, and consequently have similar organizational views. However, it is important to recognize that individuals also belong to other groups — for instance, extended families and ethnic, religious, gender, recreational and political groups — which may affect the relationships between occupational groups in various ways. For example, if a researcher and an extensionist belong to the same political party this may be an important influence on their professional interaction, transcending other influences. In short, membership of the research and extension groups cannot be looked at in isolation from membership of other groups.

Intergroup relationships can be broadly divided into two distinct but interrelated types — instrumental and expressive. Certain stakes and processes are associated with each type.

Instrumental stakes and processes. Instrumental stakes are the resources, both tangible and intangible, put at risk by each group undertaking shared activities. They reflect the commitment of each group, and its potential gains or losses in the joint venture. According to Walton, instrumental stakes are revealed when the following questions are answered:

- In intergroup plans, how much emphasis is given to the policies, programs and philosophies of each participating group?
- What proportion of each group's resources will be committed to implementing intergroup activities?
- How much operational control will each group be able to exercise over joint activities?
- How much credit or blame will each group receive in the event of success or failure? (Walton, 1973).

Walton argues that the instrumental stakes of collaborating groups may be compatible, containing integrative potential, or they may be fundamentally incompatible, in which case they will give rise to conflict. Two types of process are associated with compatible and incompatible instrumental stakes between groups, namely problem solving and bargaining. Problem solving occurs when the joint gain to both the groups is perceived as variable, while bargaining occurs when the joint gain is perceived as fixed.

Instrumental stakes and processes reflect the goals and preferences of each group and the objective conditions which facilitate and frustrate their attainment. Realistic Conflict Theory (RCT) analyzes the relationships between groups primarily in relation to group goals. RCT is the oldest and most influential theory of intergroup relationships within social psychology. It will be discussed in greater detail in the third section.

Expressive stakes and processes. Expressive stakes and processes concern the behavior of the individual that "expresses who the person and the group

---

Figure 3. Instrumental and expressive stakes and processes: Mixed motive situations

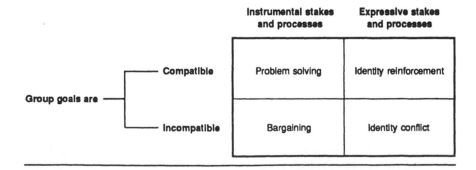

Group goals are	Instrumental stakes and processes	Expressive stakes and processes
Compatible	Problem solving	Identity reinforcement
Incompatible	Bargaining	Identity conflict

he represents wants to be in the situation and... how he perceives and feels about other participants and the group they represent" (Walton, 1973). Both individuals and groups have identity attributes, all or some of which may be at stake in their relationship. Compatible attributes lead to identity reinforcement, while incompatible ones lead to identity denial. Social Identity Theory (SIT) seeks to explain the nature of expressive stakes and processes.

Walton argues that most relationships between groups contain all four processes — problem solving, bargaining, identity reinforcement and identity denial — and, as such, are "mixed-motive" situations (*see* Figure 3). Furthermore, the prevalence of each of the four processes differs from one setting to another and can change over time.

Realistic Conflict Theory

In this section I will explore the usefulness of RCT in explaining the dynamics of relationships between groups within IATS. RCT was originally developed by Sherif and his associates in the USA during the 1950s and has subsequently formed the intellectual basis for much of the literature on inter-organizational/departmental conflict (Sherif et al., 1961).

The fundamental proposition of RCT is that intergroup attitudes and behavior are determined mainly by the nature of the goals that link groups. RCT is therefore concerned with analyzing the goals of each group and, in particular, the extent to which these groups are interdependent.

The intergroup relationship is dominated by one of two types of interdependence which RCT terms as cooperative and competitive. RCT further assumes that there is normally a clear, inverse relationship between cooperative and competitive interdependence.

Competitive Interdependence

As its name indicates, RCT takes conflict between groups as its starting point. Conflict is seen as the result of competition between groups for the same resources. This may seem an unduly narrow perspective: surely, it could be argued, there are sources of conflict other than just resources. RCT overcomes this objection by adopting a wide definition of resources, which it sees as both concrete and abstract. Thus, groups may compete for tangible resources, such as funds or physical inputs, or for less tangible ones, such as power and status which, despite their abstract nature, have direct consequences for the realization of a group's goals.

Public sector agricultural research and extension groups have a common overall objective (to improve agricultural productivity), are often part of the same organization, and are government funded. Thus, within the RCT perspective, conflicts about their goals and priorities are bound to reflect competition for both concrete and abstract resources.

Extensive research on RCT has demonstrated the powerful psychological forces which come into play between competing groups. According to this research, personality differences or previous interindividual relationships play only a limited role in influencing intergroup relationships.

Typically, open hostility escalates rapidly between groups which have a strong sense of ingroup identity and solidarity ("us and them" feelings). In such groups there is a marked overestimation of ingroup achievements and underestimation of outgroup achievements. Misperceptions and misunderstandings play an important part in the development of competition. Members of one group think that they understand perfectly the other group's position, when in fact they do not. Areas which the groups share in common are likely to go unrecognized, and to be seen as characteristic of one's own group only. Aggressive group leadership emerges.

Where competition between two groups is intense and unavoidable, RCT predicts that open hostility will break out between them. However, apex management may refuse to allow hostility to manifest itself openly, in which case the relationship will be characterized by conflictual avoidance and suppressed hostility.

In other situations, competition between groups may be low, but indifference or conflictual avoidance still dominate the relationship. Walton argues that this is mainly because of the lack of pressure to integrate activities. In such cases the costs associated with failure to solve problems or bargain appear to be very low — merely the disapproval of bureaucratic superiors, a disapproval that is frequently more apparent than real. "Failure to cooperate only involves opportunity costs, not regression from the status quo" (Walton, 1973). The factors that dissuade groups from cooperating with one another include the bureaucratic time involved, the risk that cooperation will fail, the lack of symmetrical gains in interdependent ventures, higher vulnerability to attack as a result of increased organizational visibility, and a desire to preserve a particular identity or status.

How important is competition for resources in shaping relationships between IATS groups in developing countries? I shall consider intergroup competition for the following types of resource: power and responsibility, tangible inputs, status and rewards, and impact recognition.

Power and responsibility. The ability of a group to achieve its goals depends completely on the power it possesses. Power is therefore the ultimate resource on which all other resources depend. Intergroup competition for power is likely to be important where: (1) one or both groups are unhappy with the current balance of power between them, and (2) the total amount of power available is seen as a fixed resource to be divided between the two groups; that is, the more research has, the less extension has.

In formal organizational terms, public sector agricultural research and technology transfer institutions rarely have clear-cut *de jure* power or control over each other. However, because there are other sources of power (coercive, expert, political, charismatic/personal) which either group can exploit, *de facto* power is not normally equally distributed. Apex management often favors one group over the other (or is perceived as doing so), especially when research and technology transfer are parallel branches or departments of the same ministry or parastatal. The favored group is usually agricultural research, because of its higher status and because apex managers frequently have research backgrounds, but occasionally extension may be favored by managers concerned about the limited impact of research. Policy makers may also favor extension because it has a short-run rather than a long-run time horizon, and because some extension activities may pay substantial political dividends.

In formal organizations, competition between groups for power usually takes place mainly at the leadership level. Senior IATS managers may have spent many years reaching their present positions and strong rivalries between them may have developed. Thus, the "heavy hand of the past" is likely to be much in evidence in the current competition for power. Interindividual competition among senior managers, especially for the few apex positions that offer promotion, is also an important factor. Leadership conflicts of this kind can destroy departmental cultures conducive to strong linkage relationships.

A group's power depends on the extent of its formal responsibilities. Consequently, intergroup competition for power often manifests itself as conflict over functional responsibilities. However, the desire of one group to take over some of the functions of another group may also be motivated by the conviction that the second group is not performing those functions satisfactorily. This is often the case between research and extension. In practice, then, it is rarely possible to disentangle the extent to which a group's bid for power is influenced by its attempts to stake out its access to resources in a Machiavellian sense, and its desire to fulfill what it believes are its rightful responsibilities.

Competition between groups over functional responsibilities can have a number of consequences. One or both groups may try to reduce their dependence on the other. Thus, in some countries technology transfer institutions have tried to take increasing responsibility for adaptive research. Conversely, where agricultural researchers have held negative views about the efficacy of technology transfer they have sometimes attempted to internalize various technology transfer functions (especially where these have a strong client orientation) or, alternatively, have dismissed the role of technology transfer institutions in the belief that "good technology sells itself".

The extent to which interdependence can be reduced varies according to the type of research involved, as well as the topics covered. Technology transfer institutions cannot usually undertake strategic or applied research, especially on tree crops and on crop and animal breeding. With regard to adaptive research, however, particularly in the areas of agronomy and the social sciences, there is considerable scope for expanding the role of the technology transfer group.

Tangible inputs. Competition between groups for tangible inputs is central to RCT. With regard to financial resources, the extent of competition will depend on:

- The overall availability of funds. The greater the shortage of funds in relation to present and future needs, the greater the competition is likely to be.
- The extent to which the two groups view the resources available to them as being fixed rather than variable in amount. If resources are seen as largely fixed and emanating from the same source, this will tend to encourage a competitive, bargaining type of resource procurement and allocation process.
- The degree to which the present breakdown in resources between the two groups is seen to be equitable and stable. Frequently, norms become established over time concerning this breakdown, with the preceding budget being used as the main criterion for deciding current allocations. Initiatives by apex management to change the budgetary allocations can precipitate conflict.

As part of the programming and budgeting processes, it may be useful to establish common resources that are earmarked for specific linkage activities. In this way, linkage resource requirements are clearly established, thereby reducing competition for them.

Competition between groups for human, as well as financial and physical, resources is also an important factor in many IATS. The intensity of this competition will depend on the availability of individuals who meet the required recruitment standards, and on the relative success of each group in recruiting them.

The extent to which potential recruits are attracted by the prospect of employment in each group is a critical factor. University graduates often prefer a career in agricultural research, with the result that the research group is normally able to attract recruits of relatively high calibre. Extension, on the other hand, is often regarded as less stimulating, even humdrum, and may fail to attract sufficient numbers of high-quality graduates. In addition, competitive feelings and attitudes may be intense when one group habitually "poaches" recruits from the other.

The degree of intervention by apex management in the recruitment process can also be a factor. For instance, ministries sometimes assign recruits to extension departments instead of allowing them to work in research, as part of a deliberate policy to strengthen extension. Such a policy may be resented by the research group as favoritism, thereby exacerbating competition.

Status and rewards. Intergroup competition over status and rewards is endemic in the public sector labor markets of most developing countries. Within IATS, this can often be directly attributed to the significantly higher status and rewards enjoyed by the research group. Even where rewards are broadly similar, competition may still be intense because, when negotiating increases in rewards, each group may rely heavily on making comparisons with the other. Where professional researchers have higher status than professional extensionists — as they usually do — the latter are likely to resent having to continually convince their governments that they are as valuable as researchers.

Generally speaking, there is an ascending "hierarchy of status" which goes from farmer to extensionist to researcher. According to Lionberger and Chang, this hierarchy becomes less marked as agriculture advances and becomes more specialized and sophisticated (Lionberger and Chang, 1970). Thus, one would expect competition over status and rewards to become a less serious factor in undermining linkage relationships in IATS as countries develop their agricultural sectors.

In order to assess the impact of competition for rewards and status on intergroup relationships in IATS, it would be necessary to: (1) determine the size (over time) of status and reward differences between the relevant occupational subgroups (AM, RM, EM, RP, RS, SMS and EFW, in my simple model); (2) analyze the factors responsible for the differences; (3) ascertain the extent to which the inferior subgroups (invariably those in extension) perceive themselves as disadvantaged; and (4) discover the extent to which they interact competitively with the more privileged group or subgroups in order to improve their positions. In view of the importance of this topic, it will be discussed further in the section on "Occupational Groups and Structure".

Impact recognition. Conflict between research and technology transfer groups can arise as a result of the difficulty of disentangling their specific impacts on agricultural productivity and production. When both groups are ineffective there is a strong tendency for each to blame the other. Over time this can develop into a perverse kind of interdependence whereby each group needs the other as a scapegoat.

From an RCT perspective, the resource implications of any attempt to improve the links between research and technology transfer are critical, since the benefits and costs of changes are unlikely to be the same among the two groups. It may be necessary to substantially increase the resources available to one or even both groups. The introduction of the Training and Visit (T and V) extension system, for example, is usually accompanied by budget increases for both groups. However, even this may be insufficient to overcome competition that has existed over a long period or that has been created as a direct result of the attempt to bring the two groups closer together.

In such cases it may be necessary to increase apex management control, thereby reducing, in varying degrees, the power and autonomy of one or both groups. Group, as well as apex, managers will normally be required to exercise closer supervision. Attempts to improve linkage relationships are often unpopular, and staff discontent will usually be targeted at their own group managers in addition to members of the other group. Strong leadership is needed to weather the storm.

Ultimately, if intergroup competition cannot be resolved effectively, the only solution may be to reconstitute the groups through organizational changes. This should be an option of last resort, however.

Cooperative Interdependence

Superordinate goals. RCT postulates that shared superordinate goals must be established if conflictual intergroup relationships are to be transformed into collaborative ones. A superordinate goal is one that has "a compelling appeal for members of each group, but one that neither group can achieve without the participation of the other" (Sherif et al., 1961).

In principle, IATS groups share the same superordinate policy goals, since they are both formally mandated to pursue the national agricultural objectives laid down by government. Clearly, however, in many countries exhortations for research and extension to meet the technological needs of farmers are not sufficient to ensure that they will collaborate effectively. This is often attributable to an unstated lack of government commitment to agricultural development, coupled with the inability of agricultural producers, in particular smallholder subsistence farmers, to put sufficient pressure on research and extension to ensure collaboration. Furthermore, when agricultural development policies are only broadly stated, there is considerable scope for research and extension to interpret these as they wish so that, in practice, each group's goals are not only significantly different but may, in fact, be incompatible.

Operational measures. RCT places primary emphasis on the establishment of superordinate goals, but the mere statement of such goals is not enough to ensure more integrated IATS performance. To be successful, the introduction of superordinate goals must be accompanied by operational measures, including the redefinition and redistribution of roles and tasks in such a way as to stimulate a joint problem-solving approach. The following conditions will normally have to be met:

- The technology transfer process must become central to both groups. Where necessary, other responsibilities and activities should be downgraded or eliminated in order to create the required focus on technology transfer. This has been a key aspect of the T and V approach to extension (Benor and Baxter, 1984).
- The two groups must become more interdependent in operational terms. This invariably entails important changes in the functional responsibilities of each group and therefore a significant redrawing of group boundaries. In the past, extension departments have often been too passively dependent on research for information, and generally unable or unwilling to engage in problem-solving activities. The desire to change this relationship has underpinned recent attempts by governments and donor agencies to move away from the traditional concept of extension (as "techniques, methods and means of adult education") to a new concept, that of "technology transfer", involving a major new emphasis on adaptive research and the integration of information.
- Operational goal statements and objectives should be established for both research and technology transfer groups. Wherever possible these should be objectively verifiable, and they must be strictly enforced by apex and group managements. The activities and linkage responsibilities of each group member should be clearly specified in job descriptions and, more importantly, as part of the annual programming process.
- Formal and/or informal linkage groups (or teams) comprising both research and extension personnel should be created. These would be similar in approach to task forces or to the product teams used in industrial R & D.

The leadership of new linkage groups is critical, requiring sensitivity to the different backgrounds and needs of the research and technology transfer members making up the new group. In bringing individuals from two groups together, it is important that each subgroup within the new group should continue to have distinct functions. Careful consideration must be given to identifying the appropriate positions and individuals to make up the new group. The farming systems research and extension (FSR/E) teams established recently in a number of developing countries, especially in sub-Saharan Africa, are perhaps the most concrete examples of formally constituted linkage groups of this kind. Such groups must often keep a low organizational profile during their early years, in order not to be perceived as a threat by other groups. In addition, it is important that they do not see themselves as a new *functional* group having separate aims from both research and extension.

The role of management. "Management, not the researcher, defines what it wants R & D to do" (Berman, 1973). Superordinate goals must be formulated and inculcated by policy makers and managers at both the apex and group levels. Developing political, managerial and staff commitment to superordinate goals is normally a considerable challenge. It may well be necessary to replace current management with new personnel who are not encumbered by the legacy of past tensions between the two groups. Even this drastic measure may fail if traditional institutional culture and practices are deeply ingrained at professional and subprofessional levels.

Apex management should be seen as impartial, concerned only with fulfilling client needs. This will require the elimination of any inequality in the power of the two groups fostered, consciously or unconciously, by apex management in the past. Steps should also be taken at the highest political level to encourage farmers and their organizations to take a more active role in determining the research agenda and ensuring that prescribed linkage activities are effectively undertaken.

As mentioned earlier, creating superordinate goals will normally require an increase in the supervisory control exercised by apex management, especially during the early stages of change. Nonetheless, interaction between apex and senior group managers, while becoming more intensive, should remain flexible, informal and open in order to create strong collegial relationships based on trust. There must also be some tolerance of failure among all three management groups.

The establishment of superordinate goals will normally depend on changes in the institutional cultures of research and extension. Schein defines institutional culture as "the pattern of basic assumptions that a given group has invented, discovered or developed in learning to cope with its problems of external adaptation and internal integration" (Schein, 1984). The strength of an institutional culture depends largely on the homogeneity and stability of group memberships and the length and intensity of shared experiences, as well as on leadership. Apex and group managers have the primary responsibility for establishing institutional cultures conducive to positive intergroup relationships. Normally, the greatest potential for instilling the new culture will be among recent recruits, particularly at the professional level.

The informal personal goals of both managers and staff can seriously interfere with their commitment to superordinate goals. The existence of complex networks of social obligation, coupled with pervasive patron-client relationships, are the most significant manifestations of this phenomenon. There are no easy solutions to this problem.

Donors often play key roles in formulating new superordinate goals and in the activities subsequently undertaken in pursuit of these goals. Such interventions can be useful during the early stages of an attempt to improve links, but only if they are based on a detailed understanding of the dynamics of intergroup relationships in IATS. In the past this has been the exception rather than the rule.

The commitment of staff. RCT postulates that superordinate goals can exert a positive influence only when the following conditions are met:

- The status and/or reward grievances of disadvantaged and dissatisfied groups are adequately resolved.
- Individual goals are sufficiently compatible with superordinate goals. The relative importance of individual attitudes and behavior which adversely affect the performance of formal occupational roles must therefore be taken into consideration.
- Sufficient weight is given to the attainment of superordinate goals in performance appraisal and reward systems. The conventional appraisal criteria used in agricultural research (publications, potential impact of new technology) and extension (visits/contacts, inputs distributed) often give little weight to linkage behavior.

Blake and Mouton argue that commitment to improving relationships between groups has to be based on the direct participation of group members at all levels in intensive, carefully structured interactions with one another. They believe that conventional approaches to reducing conflict (or indifference) and building cooperation are "useful" but not sufficiently "focused on penetrating the underlying dynamics" of the relationship (Blake and Mouton, 1984).

Blake and Mouton's "interface conflict-solving model" helps groups "generate the motivation and commitment required to escape an undesirable history of conflict and establish new norms of cooperation and collaboration". Their approach requires that the key members of each group to interact both with each other and with the other group, using the following steps:

Step 1: Developing the optimal relationship. Each group works separately to create a model of the optimal relationship, in the light of their own problems and needs.

Step 2: Consolidating the optimal relationship. A model of a sound relationship is then generated through the two groups' joint efforts.

Step 3: Describing the actual relationship. The current relationship is described by each group separately, with members analyzing the historical factors that have shaped it.

Step 4: Consolidating the actual relationship. The two groups' perspectives are amalgamated in a jointly agreed picture that objectively describes the current relationship.

Step 5: Planning for change. The operational changes to be made are jointly agreed upon and described in detail.

Step 6: Progress review and replanning. Dates are fixed for the two groups to reconvene 3 to 6 months after the initial session in order to review progress, criticize their new relationship, and plan the next steps.

Individuals who are "respected neutrals" should be used as group facilitators.

In most societies, people have little idea how they operate in groups in the workplace. Consequently, additional management initiatives should be regularly undertaken to increase managers' own self-awareness and that of their subordinate staff.

Healthy competition. Establishing compelling superordinate goals rarely means that all competition between groups will be eliminated. Some management specialists argue that not only is some competition inevitable within or among complex organizations but that it is also desirable. Healthy competition motivates both managers and their staff: if agricultural research and extension personnel are "too good friends" (Souder, 1980), the resulting relationship is unlikely to be sufficiently challenging. Such competition also helps ensure that difficult problems are not suppressed but dealt with quickly and openly. Where two groups have been uninterested in (or have avoided) each other, healthy competition forces them to relate, thus providing a more effective basis for the establishment of superordinate goals.

The notion of healthy competition is based on impressionistic evidence, mainly from the highly competitive environment of the USA. There is no rigorous evidence within social psychology that competition between groups is desirable in its own right. Field and laboratory studies have almost always shown that cooperative goals elicit superior performance and that therefore the overall aim of management should be to encourage as much collaboration as possible.

Social Identity Theory

Group Identity and Comparison

There has been criticizism of RCT because of its exclusive focus on group goals and competition for resources, and its corresponding lack of attention to individual motivation and the psychological processes associated with this. Thus, RCT takes the existence of groups as given and posits a simple one-to-one correspondence between the objective existence of the group and how it is perceived by group members.

SIT addresses these weaknesses of RCT by offering more strictly psychological explanations of intergroup behavior and relationships. An important difference in the application of the two approaches is that SIT has been developed primarily in situations where there is unequal power between the groups which are being studied.

With regard to links within IATS, SIT offers many useful insights. It was originally developed by Tajfel and his associates in the late 1960s in order to analyze socio-psychological processes underlying tensions between major social groups or categories, most notably ethnic, national and linguistic (Tajfel, 1982; Tajfel and Turner, 1979; Turner et al., 1979). However, SIT has also been applied to intergroup behavior in more micro-level organizational settings, and as such is directly relevant to the analysis of links within IATS (Bourhis and Hill, 1982; Brown, 1981; Brown and Williams, 1984; Brown et al., 1986).

SIT has its intellectual roots in socio-psychological theories which are concerned with the processes by which individuals interpret events. According to these theories, a fundamental human drive is the comparison of oneself with other people (Festinger, 1954 and 1957). All individuals desire to be at least as proficient as those with whom they compare themselves. They are thus motivated to test their own abilities against the abilities of others, but they will only do so in ways that are unlikely to threaten their own self-esteem. This determines the targets which are selected for comparisons, and whether an individual's feelings towards others are positive or negative.

A key proposition of SIT is that a group exists only in so far as it relates to other groups. The imperative to compare is what drives relationships between groups. SIT relies on four key concepts: social categorization, social identity, social comparison, and positive group distinctiveness.

Social categorization: Human beings are assumed to be continually active in their efforts to define themselves in relation to the world they live in. A key aspect of this basic cognitive process is social categorization, whereby each individual categorizes the social groups to which he/ she belongs and also those of other people. This segmentation of the individual's world imposes a certain order and provides the focus of self-identification.

Social identity: The individual's knowledge of his/her membership of various social categories or groups and the intellectual value and emotional significance attached to these memberships are defined as his/her social identity.

Social comparison: This is the psychological process through which characteristics of the individual's own group (the ingroup) are compared with those of other relevant groups (outgroups): "The characteristics of the individual's own group achieve most of their significance in relation to perceived differences from other groups and value connotations of these differences" (Tajfel, 1982).

Positive group distinctiveness: Individuals act in a manner to make their own groups as favorably distinctive as possible from other groups: "Every social group attempts to achieve and preserve a social identity, and such an identity is always achieved in contradistinction to an outgroup" (Tajfel and Turner, 1979). It follows, crucially, that "an ingroup may discriminate against outgroups not because there is any realistic conflict of group interests but simply to differentiate themselves and maintain a positive social identity for their members". In other words, the process of comparison creates competitive relationships, and this is as much a subjective, psychological process as it is the result of objective conditions.

If a group has "an inadequate social identity", its members will, according to SIT, attempt to change this situation in a number of ways. If the inferior status is generally accepted, members will seek to achieve a positive self-image by individualistic means, notably by making favorable comparisons with other ingroup members, thereby negatively affecting ingroup cohesion, or by trying to gain entry into the higher-status outgroup. If, however, members reject their lower status, SIT identifies three possible strategies for group action: (1) gain equality with the outgroup on relevant characteristics; (2) redefine negatively valued characteristics positively (the "black is beautiful" counter-response); (3) create new dimensions not previously used in intergroup comparisons. Based on empirical analysis of the processes of group categorization and comparison, SIT theorists have developed a number of general propositions:

- The stronger the importance attached to belonging to a group, the stronger ingroup favoritism and outgroup prejudice will be. In other words, the level of conflict between groups is likely to be that much greater the more group members are committed and loyal to their own group and the less interested they are in the goals of other groups.
- During intergroup conflict, group cohesiveness increases because the external threat to the group brings members closer together.
- Stereotyping, based on the conviction that "they are all the same", is a widespread symptom of conflict between groups.
- There is a universal tendency to exaggerate the differences between individuals falling into different categories and to minimize these within categories (Tajfel's law of categorization).
- The more similar groups are in terms of a characteristic, the greater the need to create positive group distinctiveness with regard to this characteristic and the more likely, therefore, that mutually competitive behavior will occur.
- The more threatened the ingroup by the outgroup, the greater the ingroup antipathy towards the outgroup.
- Disadvantaged/dominated group members define themselves more in terms of their social position and their group membership, whereas more advantaged group members conceive of themselves less in terms of the group and more in terms of personal characteristics.
- An ingroup can maintain positive social identity by avoiding unfavorable comparisons with more advantaged outgroups.
- The more distinctive a particular group attribute, the more it will tend to be evaluated positively.
- It is not possible to predict with any certainty what strategies a group will adopt in order to improve its identity.

The Relevance of Social Identity Theory

Although SIT is more abstract and theoretically complex than RCT, it provides a highly relevant framework within which to analyze relationships between groups within IATS. SIT nicely complements the neo-Weberian theories of occupational structure and interaction, described later in this paper. On the limited evidence available, the processes of categorization and comparison identified in SIT seem universal, not specific to certain cultures. And the analytical methods of SIT, while requiring careful, structured questioning of respondents, are relatively straightforward.

In accordance with the key concepts outlined above, a SIT analysis of relationships between groups in IATS would seek to: (1) unravel the process of group categorization undertaken by individuals engaged in research and extension (including apex management); (2) ascertain the relative importance attached to ingroup memberships; (3) evaluate positively valued group characteristics and the level of group distinctiveness; and (4) identify the strategies used by individuals or groups to rectify negative social identity. This latter part of the analysis would be particularly relevant to extension personnel, who often suffer from negative social identity.

SIT is more pessimistic and thus more conservative than RCT on the potential role of management in improving intergroup relationships, at least in the short term. This is because comparisons between groups are usually the outcome of a long history of deep-seated status differences between them. Nor does SIT claim to be able to predict the likely strategy of a disadvantaged group. Thus, in practical terms, SIT tends to place most emphasis on changing group structures as the only effective way of breaking down disfunctional group categorizations and comparisons.

With regard to IATS, this implies abandoning or radically changing the current division of labor between "research" and "extension" staff. This is perhaps most feasible in small countries, where there are fewer and more closely knit staff. For example, in Jordan research and extension as separate job categories have been replaced by production specialists. The more ambitious FSR and on-farm research programs may also be radical enough to trigger the reconstitution of group identities. But it remains to be seen whether this will also be true of T and V extension systems, which usually do not fundamentally change the job structure of extension or research.

Intragroup Characteristics and Intergroup Contact

In this section I will consider two sets of factors which, although they are not directly covered by RCT and SIT, can influence the relationships between IATS groups. The first set of factors concerns *intra*group characteristics; these are work orientation and style, competence, and group size, complexity and cohesiveness, while the second set focuses on intergroup contact and communication.

Intragroup Characteristics

RCT emphasizes those characteristics of groups that are directly related to competition for resources. SIT, on the other hand, being concerned with group identification and comparison, focuses more on group characteristics as explanatory factors in their own right. Nevertheless, it is usually not possible to determine a priori how specific similarities and differences between groups will affect their relationships. In general terms, both RCT and SIT predict that the greater the similarity between groups, the greater the likelihood of conflict. However, it is usually the differences, not the similarities, that are seen by the group members themselves as the cause of conflict. These differences are most evident in the areas of work orientation and style.

Work orientation and style. Research and extension are distinct functions with distinctive tasks and other characteristics. Thus, regardless of resource competition and/or group identity issues, each group is "bound to see things differently" (Thomas et al., 1972). This leads to legitimate differences of opinion between the two groups. When these get out of hand, intergroup relationships become openly hostile.

The alleged differences, in developed countries, between R & D personnel and those involved in operations (mainly marketing and production) have been extensively discussed in the R & D literature. The most commonly cited differences are listed here:

- "Researchers primarily identify with their profession, not their employer. They have no natural affiliation with industry" (Biller and Shanley, 1975). Consequently, researchers keep themselves separate and isolated from the production process.
- The goals of researchers are broader, less precise and measurable than those of operational personnel: "R & D are mostly concerned with big fundamental changes, whereas operations are happy with small incremental ones" (Westwood, 1984).
- Researchers have a longer time perspective, whereas operational personnel are constantly engaged in resolving immediate problems. The latter therefore tend to perceive their environment as being more uncertain, and are more dependent on the activities of other colleagues.
- There is more informality and collegiality among researchers than among operational personnel, and a general aversion to bureaucracy.

Similar differences between researchers and extensionists are thought to exist in developing countries. However, hard evidence to support the existence of these differences in both developing and developed countries is lacking. Indeed, where alleged differences have been systematically investigated, the findings have often been inconclusive. In a comprehensive survey of industrial corporations in the USA, Gupta and his associates concluded that "the widely held notion that R & D and marketing managers are simply different and thus cannot cooperate with each other appears to have little substance" (Gupta et al., 1986).

Competence. The competence of a group in performing prescribed linkage activities is clearly a critical factor. Within the extension group, the main skill deficiencies are likely to center on the ability to gather and disseminate information and to package and test technologies. Among the research group, communication and training skills (rather than technical/intellectual competence per se) usually give the greater cause for concern.

In order to interact as equal partners with researchers, technology transfer personnel responsible for linkage activities should have similar levels of technical competence to those of researchers. As the agricultural sector develops, ordinary extension field workers too will experience pressure to upgrade their abilities as their role as simple information disseminators becomes instead one of knowledge integrator and farm advisor.

Group size, complexity and cohesiveness. There is little firm evidence concerning the effects of group size, complexity and cohesiveness on relationships between groups. It is sometimes argued that the larger and/ or more complex the group, the more disputes it is likely to have with other groups. This is because it is more difficult to achieve a consistent point of view among group members; inconsistency confuses the members of other groups, putting a strain on relationships with them (Zander, 1982). However, the opposite argument could also be true: since a smaller group will find it easier to maintain cohesion, it might develop stronger prejudices against other groups.

It has also been suggested that groups which have coped successfully with serious intragroup tensions and conflicts are less likely to develop strong feelings against other groups (Billig, 1976). Again, the opposite might also be true: group leaders might exacerbate intergroup tensions to distract attention from current intragroup problems.

Intergroup Contact and Communication

The level of contact between groups has an important influence on their relationships. Often, contact between IATS groups in developing countries is relatively limited. There are several reasons for this: (1) the spatial separation of research and extension personnel across different geographical locations, a factor often exacerbated by poor transport and telecommunication links; (2) inadequate mechanisms for bringing together research and extension personnel to discuss issues of mutual interest; (3) high rates of staff attrition, thwarting the development of close relationships between individuals in each group, especially among those who occupy key linkage positions. In many IATS, therefore, poor intergroup communication is simply the consequence of the relevant groups being "out of sight, out of mind". The implication is that all contact is beneficial and should be encouraged. However, RCT argues that if contact between groups takes place in the absence of compelling superordinate goals, it may serve only as a medium for further accusations.

Contact theories. These theories are generally based on the belief that direct interaction between individuals from different groups builds up mutual understanding and reduces tensions. The theories have been tested mainly among ethnic groups with deeply ingrained prejudices (Hewstone and Brown, 1986). Such prejudices are unlikely to exist between occupational groups within IATS. However, contact theories highlight the need to think carefully about how contact between groups should be structured. Cook predicts that less derogatory outgroup attitudes will result when individuals have personal contact with members of a group they dislike under the following four conditions:

- participants from the two groups have equal status within the confines of the contact situation
- the characteristics of outgroup members with whom there is contact refute the prevailing outgroup stereotype (that is, the outgroup members are seen as atypical)
- the contact situation has high "acquaintance potential" (that is, it enables individuals to get to know each other as individuals, rather than as group members)
- the social norms within and surrounding the contact situation favor "group equality" and "egalitarian intergroup association" (Cook, 1978)

There is not complete agreement among contact theorists concerning these four conditions. Brown and Hewstone argue that unless contact personnel are viewed as typical of the outgroup, there will be no generalization to the rest of the outgroup — the contact personnel will be seen as the exception that proves the rule. They propose that the contact situation should be such as to encourage participants to interact with each other as members of their respective groups. The trick is then to ensure that the positive attributes of each group's stereotype of the other are emphasized (Brown, pers. comm.).

Another aspect of contact theory concerns contact initiation. It has been suggested that if one group predominates in initiating contacts with another group, the latter tends to become irritated and resentful (Whyte, 1957). This may be an important factor influencing relationships between IATS groups, especially where the research-to-extension communication function is dominant and extension has an essentially passive role in the technology generation process.

R & D communication processes. "The overall pattern of the innovation process can be thought of as a complex net of communication paths linking the various stages of the process" (Rothwell and Robertson, 1973). However, sociometric studies of research organizations in advanced industrial countries have consistently found that:

- Formal organizations create serious barriers to effective communication with external sources of information.
- "The technologist cannot communicate well with outsiders" (Allen, 1971).
- A high proportion of the communication that does take place is undertaken informally. It is therefore not possible to design a pattern of communication based wholly on a particular organizational structure and/or set of job positions.
- A fairly small number of individuals is responsible for a high proportion of internal and external communication. These "gatekeepers" evolve to fulfill a need the organization itself cannot satisfy (Allen et al., 1979).
- Given group and intergroup goals, there is an optimal flow of communication. Trying to establish more intensive communication beyond this level is disfunctional and leads ultimately to chaos.

Detailed sociometric research on the communication networks of IATS in developing countries has not yet been undertaken. However, many of the findings are likely to be the same as those in developed countries.

Certainly, it is highly probable that IATS contain individuals who can be regarded as gatekeepers. Management should focus on identifying actual or potential gatekeepers in each group and, where appropriate, should provide training or other kinds of assistance to help them fulfill their communication roles more effectively.

Other, more obvious ways of improving contact and communication in IATS center on intensifying the level of personal contacts between groups. On a day-to-day level this could be achieved by increasing the physical proximity of groups by, for example, locating group members in the same office buildings. Regular meetings, including joint workshops and seminars, are of paramount importance. In some situations, contact can be facilitated by the temporary secondment of individuals from one group to another. This has been tried by some IATS for young, newly recruited professional staff, as part of their apprenticeship training.

Occupational Groups and Structure

As mentioned earlier, the distribution of status and rewards is critical in influencing relationships between groups within IATS. In order to identify the determinants of group status and rewards, however, it is necessary to go beyond the socio-psychological explanations of intergroup relationships that have been the focus of attention so far, and briefly consider more mainstream sociological theories which help to explain the occupational structure of a society. As with other related groups of occupations (such as health, law and engineering), especially in the service sectors, the occupational structure of IATS is powerfully influenced by the constitution of their professional groups and, in particular, by the relationships of these groups with each other and with non-professional occupational groups.

Neo-Weberian Theory

The traditional definition of a profession is that its practitioners acquire a set of skills which are, in some way, superior to those skills possessed by other individuals: "It is the existence of specialized techniques acquired as a result of prolonged training which gives rise to professionalism and accounts for its peculiar features" (Carr-Saunders and Wilson, 1964). Adherence to certain ethical codes and the overall regulation of work practices by members are the other commonly cited characteristics of a profession.

This approach to the professions has two main problems. First, it is ahistorical: little attempt is made to investigate the profession's evolution. Such an investigation would focus on the processes whereby the profession establishes its place in society and its response to pressures for change, such as technology developments, government policies and practices, and shifts in economic and social power. Second, this approach uncritically accepts the profession's own definition of itself, which focuses on its perceived role and contribution to society. Only those aspects of professional behavior that conform with the profession's own ideology are emphasized.

Instead of the traditional approach, I propose to adopt a neo-Weberian perspective in analyzing occupational relationships in IATS. The basic proposition of the neo-Weberian theory of occupational structure is that occupational groups compete with one another for status and economic rewards. The main form this competition takes is the attempt by groups to "inclose" themselves, excluding rival groups. Through the labor market, bargaining occurs between individuals with different skills and formal qualifications which constitute their "market capacity".

The principal means by which occupational groups preserve or attempt to improve upon their position in the hierarchy is through the erection of "barriers to entry" based mainly on the acquisition of academic qualifications which are sufficiently superior in content and length of training to create the desired distance between themselves and rival groups. This emphasis on credentials is likely to become acute at times when education and training are expanding, as occupational groups attempt to escalate the qualification barriers between themselves and their increasingly well-educated rivals further down the hierarchy (Dore, 1976).

The neo-Weberian approach focuses mainly on the process of professionalization. Professionalization is the development over time of the market capacity of an occupation, which eventually enables it to establish a monopoly over specific areas of the division of labor. Originally, a profession could attain and maintain its position "by virtue of the protection and patronage of some elite segment of society which had been persuaded that there was some special value in its work" (Friedson, 1972). In time, the role of patron and protector may be gradually assumed by the state, especially when the occupation is employed mainly in the public sector, as is often the case for agricultural research and technology transfer in developing countries.

Professionalization comprises the processes of inclusion and exclusion that characterize the development of a professional group. Successful "inclosure" depends on the ability of a profession to establish control over two key processes; namely, the creation of an exclusive protected market for its skills, and the education and training of new recruits into the profession.

To be marketable, the "professional commodity" has to be distinct and recognizable to clients and the public at large. This necessitates the establishment of an intellectual basis for the profession. Hence the creation and regulation of a market of professional skills hinges on the ability of the profession to control the production of the professionals themselves (that is, to control the relevant part of the formal education and training system). "The singular characteristic of professional power is that the profession has the exclusive privilege of defining both the content of its knowledge and the legitimate conditions of access to it, while the unequal distribution of knowledge protects and enhances this power" (Friedson, 1972).

Status and Income Differences

Marked differences in occupational status and income between professional and subprofessional personnel are a common feature of the occupational structure of many developing countries. This is especially true of groups within IATS.

The main factor responsible for these differences is that an individual's occupational position in the modern sector of most developing countries continues to be overwhelmingly dependent on educational attainment. (This is also true in developed countries, but the importance of education and qualifications is generally less pronounced). In determining public sector grades and remuneration, considerably less importance is attached to experience and skills gained on the job by subprofessional personnel, compared with equivalent positions in Europe and North America.

Another contributory factor in much of Africa and, to a lesser extent, Asia, is the continuing legacy of colonial public sector grade and salary structures. Government jobs were racially segregated during the colonial period; Europeans occupied professional positions and were remunerated on the basis of metropolitan salary structures.

Nationals, on the other hand, were generally confined to subprofessional positions and were paid on the basis of prevailing local labor market conditions. In many developing countries political independence did not result in a major reform of the inherited colonial grade and salary structures. Nationals took over the jobs of the European colonizers, but there was little desire on the part of the new political and bureaucratic elites to undermine the privileged economic and social position of themselves and their colleagues. Although successive government financial crises, coupled with the increasing supply of trained personnel, have gradually eroded income differences, they remain considerable in many countries (Bennell, 1982).

In more general terms, the marked separation of professional and subprofessional staff in many developing countries is symptomatic of the greater dominance of professional groups in the class structure of these societies. This dominance has inhibited the emergence of occupational structures conducive to meeting the needs of predominantly poor and rural societies (Bennell, 1982; Bennell, 1983). It is against this background that relationships between IATS groups must be examined.

Agriculture Research: An Established Professional Group

From their inception, agricultural research organizations in developing countries have generally enjoyed relatively stable occupational structures, with few pressures for change. In large part this can be attributed to the fact that, as research organizations, they have been labelled "professional", a status originally acquired through the transfer of professional researchers from the colonial countries. The emergence of a local agricultural research establishment is a fairly recent development. As a result, the professional status of the agricultural researcher in most developing countries has never been seriously questioned.

Against this background, the professional "inclosure" of the agricultural researcher has generally been highly effective. The heightened occupational separation between professional and subprofessional personnel has meant that agricultural researchers in developing countries have not, in general, felt threatened by support staff seeking significant restructuring of the occupational division of labor.

Nevertheless, agricultural researchers have frequently had to struggle to maintain or improve their status in comparison with higher-status, more established professions, such as medicine, law and engineering. In comparison with these professions, agricultural researchers are a relatively small group whose close association with farmers and agriculture has not helped their image. In addition, their bargaining power has traditionally been weak, often reflecting the lower priority afforded by government to rural as opposed to urban development.

In order to promote their group identity, agricultural researchers have generally tried to ensure that their market capacities remain at least equal to those of other professions. This has encouraged greater specialization and sophistication, often associated with raising qualification requirements to the post-graduate level (of course, there have also been genuine technical reasons for increased specialization and training). Equally important has been the corresponding need to distance the profession from lower-status professions and occupations, in particular, with which it tends to be most commonly associated, namely agricultural extension. This may involve attempts to maintain the subordinate status and income positions of extension workers which, for obvious reasons, is likely to engender conflict.

There are exceptions to this picture. In some countries researchers do not perceive themselves as a separate professional group, or at least attach limited importance to the difference. In Taiwan in the 1960s, for example, it was observed that "agricultural technicians in research institutes and improvement stations do both extension and research and regard themselves generally as being part of both. They rarely regard themselves as exclusively one or the other" (Lionberger and Chang, 1970).

Status distinctions among the various disciplines within agricultural research may also affect intergroup relationships in IATS. However, the precise nature of these effects remains unclear. According to the propositions of SIT, lower-status disciplines (typically social scientists and general agronomists) with insecure social identities should be less likely to cooperate with technology transfer groups, which have a lower status still. On the other hand, perhaps the belief systems and work orientation of such disciplines make them more inclined than other disciplines to cooperate with technology transfer groups.

Agricultural Technology Transfer: An Emerging Professional Group

In many developing countries, public sector technology transfer activities were not formally institutionalized until after the establishment of agricultural research organizations. As latecomers, technology transfer personnel experienced difficulty in gaining credibility with other public sector agricultural personnel, in particular agricultural researchers.

A key characteristic of agricultural technology transfer organizations during this early period, particularly in Africa and Asia, was that they were staffed mainly by subprofessionals. In part this reflected the relative underdevelopment or novelty of public sector agricultural extension in the colonial countries, in contrast to agricultural research. More important, however, was the narrow concept of technology transfer, in which the extensionist was for the most part seen merely as a disseminator of information to peasant farmers, concerned to "show how" rather than to "know how". That such a concept prevailed was not surprising given the huge size of farming populations and the chronic shortage of skilled staff. Moreover, extension workers had to live near their farmer clients, and only relatively low-status, non-professional employees could be expected to work in these rural environments, which at that time generally lacked basic amenities. Thus, in many countries there were few professional extensionists or even well-trained subprofessional extension workers who could have acted as effective intermediaries between research scientists and field-level extension workers.

Given this historical legacy, links within IATS were, and in many countries still are, characterized by a professional research-subprofessional extension relationship. The distinction between mental and manual labor is particularly important here. Research scientists are seen as the embodiment of a high level of mental labor, while the extension workers' role is often seen as involving low-level mental activities ("delivering messages") and manual activities (delivering inputs and preparing demonstration plots).

The professional research-subprofessional extension relationship has, more often than not, resulted in tensions between the two groups, mainly because researchers have tended to adopt patronizing attitudes towards extensionists who, not surprisingly, have resented being treated in this manner.

Extension's close association with small farmers has further served to undermine its status, during a period when governments have been generally preoccupied with industrialization. Nor, except in a few countries with more advanced agricultural sectors, have there been any significant numbers of people engaged in technology transfer activities in the private sector, whose higher status and pay could have served as a powerful reference point for their public sector counterparts.

With their relatively low status and poorer pay and working conditions, agricultural extension services have generally been unable to attract good quality recruits. Research has invariably been the preferred career choice among university graduates in agriculture. Some governments have responded by compelling graduates and others to take up jobs in extension, but this has merely increased resentment and job dissatisfaction. In many countries, particularly in Asia (including Nepal, Philippines, Sri Lanka, Taiwan and South Korea), it appears that extension recruitment standards have continued to deteriorate. In South Korea, 48% of extension agents were university graduates in 1965; this had fallen to 25% by the late 1970s (Asian Productivity Organization, 1980). Many agricultural extension organizations have found it difficult to retain their more able and experienced senior staff, especially where private sector employment opportunities have been relatively plentiful. As a result of high levels of attrition, the average age of extension staff is often relatively young, making it more difficult for them to build relationships with older research personnel.

The inferior status of agricultural extension staff has made the occupational structure of extension organizations in most developing countries inherently unstable, as extension personnel have tried to improve matters. Only a small proportion have been able to overcome their negative social identity by joining research organizations. The other individualistic response postulated by SIT (making favorable comparisons with other ingroup members) may have occurred in some developing countries, with professional extensionists trying to distance themselves from subprofessional junior colleagues.

However, the most common response by extension staff has been to try to improve their position through professionalization. On the one hand, this has been justified in strictly functional and technical terms, given: (1) the increasing sophistication of farmers and the agricultural sector in general, and the corresponding need to upgrade extension workers' skills; (2) recent recognition of the importance of certain functions and activities (in particular, information integration and adaptive research) which hitherto have not been adequately performed by research or extension; and (3) the greater availability of skilled personnel. On the other hand, professionalization is also a social process, enabling extensionists to attempt to increase their market capacities and thus their status and incomes. If the attempt is successful it should, according to both RCT and SIT, create the necessary sociopsychological conditions for more effective intergroup relationships.

The professionalization of extension has taken different forms according to local conditions. In many countries, the introduction of T and V marked the beginning of the process; it is likely to be a slow one, passing through several different stages. The characteristics of the new professional group have to be formalized and the necessary government bodies convinced of the need for change. The impact on government funding, particularly where the technology transfer subsystem is large, as in many Asian countries, is a critical issue. Professionalizing extension will increase salary costs. Hence the pace of professionalization in most countries will be slow.

Generally, the new professional extension service has begun by covering technical service functions undertaken mainly by subject-matter specialists (including communication and training experts); gradually, it has extended its activities to include field service functions. Extension has needed to take on significant additional responsibilities and functions in order for its managers to make a convincing case for the professionalization of their staff.

The response of public sector research organizations to the professionalization of extension has varied. To the extent that research has felt the need to distance itself from extension, attempts by the latter to close the occupational gap have worsened tensions between the two groups. Typically, disputes have centered on the involvement of extension in adaptive research and knowledge integration.

A research department can respond in several ways. First, it can respond cooperatively, accepting the need for greater involvement of extension in adaptive research and seeking to integrate the activities of the two departments as prescribed. Second, it can respond evasively, by redefining its mandate and sphere of knowledge, often by increasing the sophistication of research which, with greater emphasis on postgraduate training, forms the new basis for group distinctiveness. In other words, researchers give up responsibility for adaptive research, which in most countries they regard as lower status anyway. Thus, they maintain their separate professional identity and subvert official attempts to establish closer research-extension links. This response is most likely to occur in larger IATS, especially those that have traditionally shown little interest in adaptive research in the first place.

The third response is to engage in conflict. This is common where adaptive research forms the major part of the mandate of the research department/subsystem, as it does in most developing countries with small IATS. Disputes under these conditions can be serious: the research subsystem's *raison d'être* (at least officially) is to undertake adaptive research; thus any attempt to increase the involvement of extension threatens its professional and institutional identity. This has occurred in some countries where research has been weak. Whether organizational arrangements such as FSR/E teams are effective in overcoming conflicts of this kind remains to be seen.

In some South American countries (Argentina, Chile, Colombia, Venezuela and Uruguay), recognition of the professional role of extensionists occurred much earlier — even during the establishment phase of the relevant public sector institutions. This was mainly because of the relative sophistication of the agricultural sector and the greater availability of graduates. Another important factor is the part played by US aid agencies and universities in setting up national IATS, given that extensionists had achieved full professional parity in the USA by the early 1950s. The professional identity of the South American *ingeniero agrónomo* has also been strengthened by the introduction of compulsory registration and the development of national professional associations for agriculturalists.

Although the differences between research and extension have been less pronounced in South America, linkage relationships are still poor. From an RCT perspective, this is attributable to deep-seated competition between the two groups. For SIT, on the other hand, it is precisely the similarity of the two groups (in terms of background, age, qualification, etc), in addition to power differences between them, which is likely to have been the major cause of tension.

Another possible contributory factor has been the growing numbers of extensionists and researchers employed in the private sector, whose higher levels of pay have posed an increasingly serious threat to the viability of public sector research and extension. Faced with this challenge, intergroup professional relationships in a number of countries have been placed under increasing pressure as the morale and confidence of staff have declined.

Conclusion

The purpose of this paper has been to explore the relevance of socio-psychological and sociological theories in explaining relationships between IATS groups in developing countries. It is clear that each of the three main sets of theories that have been discussed, namely RCT, SIT and neo-Weberian occupational theory, does provide important insights into the underlying nature of these relationships. The role, and thus the relative importance, of strictly individual factors is, however, much less clear. Further theoretical investigation in this area is therefore required.

RCT and SIT offer sufficiently different explanations of intergroup relationships for them to be regarded as competing theories whose validity should be subject to further empirical investigation. SIT and neo-Weberian theory are highly complementary and together constitute a powerful cross-disciplinary model which, in many respects, is more comprehensive and incisive than RCT's exclusive preoccupation with group goals. It is precisely what lies behind these goals that has to be unravelled.

The exploratory power of these theories has been demonstrated. The next step should be to carry out carefully selected case studies of IATS that would seek empirical evidence of the role of strictly socio-psychological factors. As with all social science research, however, neatly separating out the influence of socio-psychological factors from explanations provided by other disciplinary areas will remain problematic.

Acknowledgments

The author is indebted to Dr Rupert Brown, Senior Lecturer, Institute of Social and Applied Psychology, University of Kent, for his assistance in the preparation of this paper. ISNAR colleagues David Kaimowitz, Deborah Merrill-Sands, Ajibola Taylor and Larry Zuidema also made valuable comments. Any shortcomings in this paper, however, are the responsibility of the author.

References

Akinbode, A. "An analysis of interorganizational relationships of agricultural research teaching and extension in Western Nigeria." Dissertation. Madison: University of Wisconsin, 1974.

Allen, T.J. "Communication networks in R & D laboratories." *R & D Management* (1, 1971).

Allen, T.J., Tushman, M.L., and Lee, D.M.S. "Technology transfer as a position in the spectrum through development to technical services." *Academy of Management Journal* (22, 1979).

Asian Productivity Organization. "Farmer education and extension services in selected Asian countries." Tokyo: APO, 1980.

Balaguru, T., and Rajagopalan, M. "Management of agricultural research projects in India. Part 2: Research productivity, reporting and communication." *Agricultural Administration* (32, 1986).

Bennell, P.S. "The colonial legacy of salary structures in anglophone Africa." *Journal of Modern African Studies* (20, 1982).

Bennell, P.S. "The professions in Africa: A case study of the engineering profession in Kenya." *Development and Change* (14, 1983).

Benor, D., and Baxter, M. *Training and Visit Extension.* Washington D.C.: World Bank, 1984.

Berman, S.I. "Integrating the R & D department into the business team." *Research Management* (July, 1973).

Biller, A.D., and Shanley, E.S. "Understanding the conflicts between R & D and other groups." *Research Management* (1975).

Billig, M.G. *Social Psychology and Intergroup Relations.* London: Academic Press, 1976.

Blake, R.R., and Mouton, J.S. *Solving Costly Organizational Conflicts.* San Francisco: Jossey Bass, 1984.

Bourhis, R., and Hill, P. "Intergroup perceptions in British higher education: A field study. "*Social Identity and Intergroup Relations.* Tajfel, H. (ed). Cambridge: Cambridge University Press, 1982.

Brown, R.J. "Divided we fall: An analysis of relations between sections of a factory workforce." *Social Identity and Intergroup Relations.* Tajfel, H. (ed). Cambridge: Cambridge University Press, 1981.

Brown, R.J., Condon, S., Mathews, A., Wade, G., and Williams, J. "Explaining intergroup differentiation in an industrial organization." *Journal of Occupational Psychology* (59, 1986).

Brown, R.J., and Williams, J. "Group identification: The same thing to all people? " *Human Relations* (51, 1984).

Carr-Saunders, A.M., and Wilson, P.A. *The Professions.* London: Frank Cass, 1964.

Cernea, M.M., Coulter, J.K., and Russell, J.F.A. *Research-Extension-Farmer: A Two-Way Continuum for Agricultural Development.* Washington D.C.: World Bank, 1984.

Compton, J.L. "The integration of research and extension." *The Transformation of International Agricultural Research and Development.* Compton, J.L. (ed). Boulder: Lynne Rienner, 1989.

Cook, S.W. "Interpersonal and attitudinal outcomes in cooperating interracial groups." *Journal of Research and Development in Education* (12, 1978).

Dore, R. *The Diploma Disease: Education, Qualification and Development.* London: Allen Unwin, 1976.

Festinger, L. "A theory of social comparison processes." *Human Relations* (7, 1954).

Festinger, L. *A Theory of Cognitive Dissonance.* Stanford: Stanford University Press, 1957.

Food and Agricultural Organization. *Proceedings of World Conference on Agricultural Education and Training.* Rome: FAO (with UNESCO and ILO), 1979.

Friedson, E. *The Medical Profession: A Study of the Sociology of Applied Knowledge.* New York: Dodd Mead, 1972.

Frosch, R.A. "R & D choices and technology transfer." *Research Management* (May, 1984).

Gupta, A.K., Raj, S.P., and Wilemon, D. "R & D and marketing managers in high-tech companies: Are they different?" *IEEE Transactions on Engineering Management* (EM-33, 1986).

Hewstone, M., and Brown, R. "Contact is not enough: An intergroup perspective on the contact hypothesis." *Contact and Conflict in Intergroup Encounters.* Hewstone, M., and Brown, R. (eds). Oxford: Blackwell, 1986.

Leonard, D.K. *Reaching the Peasant Farmer: Organization Theory and Practice in Kenya.* Chicago: University of Chicago Press, 1977.

Lionberger, H.F. "Functional requirements for agricultural research-extension systems: A mix of R & D and IPPS sub-systems." *Journal of Extension Systems* (1, 1986).

Lionberger, H.F., and Chang, H.C. *Farm Information for Modernizing Agriculture: The Taiwan System.* New York: Praeger Press, 1970.

Palmer, A.F.E., Violic, A.D., and Kocher, F. "CIMMYT's approach to the relationship between research and extension services and the mutuality of their interests in agricultural development." *More Food from Better Technology.* Rome: FAO, 1983.

Rothwell, R., and Robertson, A.B. "The role of communications in technological innovations." *Research Policy* (2, 1973).

Schein, E.H. "Coming to a new awareness of organizational culture." *Sloan Management Review* (Winter, 1984).

Sherif, M. *Group Conflict and Cooperation: Their Social Psychology.* London: Routledge and Kegan Paul, 1966.

Sherif, M., Harvey, O.J., White, B.J., and Hood, W.R. "Intergroup conflict and cooperation: The Robbers' Cave Experiment." Norman, Oklahoma: University of Oklahoma Book Exchange, 1961.

Sigman, V.A., and Swanson, B.F. "Problems facing national agricultural extension in developing countries." Champaign-Urbana: University of Illinois, 1982.

Souder, W.F. "Promoting an effective R & D marketing interface." *Research Management* (July, 1980).

Swanson, B. F., and Rossi, J. "International directory of national extension systems." Champaign-Urbana: University of Illinois, 1981.

Tajfel, H. (ed). *Social Identity and Intergroup Relations.* Cambridge: Cambridge University Press, 1982.

Tajfel, H., and Turner, J.C. "An integrative theory of intergroup conflict." *The Social Psychology of Intergroup Relations.* Austin, W.G., and Worchel, S. (eds). Monterey: Brooks-Cole, 1979.

Tannenbaum, A.S. "Organizational psychology." *The Handbook of Social Psychology.* Lindsay, G., and Aronson, E. (eds). New York: Random House, 1985.

Thomas, K.W., Walton, R.E., and Dutton, J.M. "Determinants of interdepartment conflict." *Inter-organizational Decision Making.* Tuite, M.F., Chisolm, R.K., and Radnor, M. (eds). Chicago: Aldine, 1972.

Turner, J.C. "Towards a cognitive redefinition of the social group." *Social Identity and Intergroup Relations.* Tajfel, H. (ed). Cambridge: Cambridge University Press, 1982.

Turner, J.C., Brown, R.J., and Tajfel, H. "Social comparison and group interest in group favouritism." *European Journal of Social Psychology* (9, 1979).

Walton, R.E. "Interorganization decision making and identity conflict." *Interorganizational Decision Making.* Tuite, M.F., Chisolm, R.K., and Radnor, M. (eds). Chicago: Aldine, 1972.

Westwood, A.R.C. "R & D linkages in a multi-industry organization." *Research Management* (May-June, 1984).

Whyte, W.H. *The Organization Man.* London: Jonathan Cape, 1957.

Zander, A. *Making Groups Effective.* San Francisco: Jossey Bass, 1982.

5

Links between On-Farm Research and Extension in Nine Countries

Peter Ewell

In most developing countries, agricultural research and extension are separate public institutions with different mandates and different ways of operating. Even where they are formally located in the same ministry, they usually have very different organizational structures and operational procedures.

The predominant model for the generation and transfer of agricultural technology is based at least implicitly on systems for breeding, testing, and distributing improved crop varieties. Researchers are expected to develop superior genetic material and/or production techniques, which they then turn over to extension for demonstration and diffusion to farmers.

Top-down systems of this kind have functioned reasonably well to meet the demands of resource-rich farmers, as well as those of both large- and small-scale producers of high-value commodities. These farmers have been able to communicate their needs to researchers, either directly or through producers' organizations, and to assess and adapt the recommendations which come to them through the extension system.

However, the lack of effective links between research and extension institutions has impeded the development and transfer of technology appropriate for small-scale, resource-poor farmers, particularly those in low-potential, heterogeneous agro-ecological areas. These farmers have no effective organizations through which to make their needs known.

Researchers do not receive enough information about these farmers' conditions and resources to set relevant priorities and goals. At the same time, local extension agents do not receive the information and cooperation they need to first adapt and then diffuse appropriate technology. The lack of good communication between research and extension has particularly limited the transfer of technologies other than improved crop varieties, such as storage and pest management methods. Rather than improved inputs, which are physically distributed, these often consist of concepts which must be reinterpreted and adapted to each new situation (Horton, 1986; Rhoades, 1987).

Extensionists are caught in the middle in many ways. They are often responsible for a broad range of government services in rural areas, of which technology transfer is only one. Seldom do they receive adequate resources for field work and travel. They are obliged to promote whatever technology comes down to them, even if it is not adapted to local agro-ecological or socio-economic conditions. They are almost always separated from researchers by wide gaps in educational level, status, salaries and social class. Researchers blame them for their failure to transfer innovations which have shown promise under experimental conditions, and for their apparent inability to provide systematic feedback. Farmers often see them as incapable of providing answers to local problems and needs (Collinson, 1985).

Farming systems research (FSR), and especially on-farm research, has been promoted as a way of developing appropriate technology and adapting it to the specific agro-ecological and socio-economic conditions of small-scale farmers. Many national agricultural research systems have developed interdisciplinary programs of this kind, with two major objectives:

1. To diagnose needs and constraints at the farm level
2. To adapt technologies to the agro-climatic and socio-economic conditions of target producers

Parallel initiatives within extension institutions have also been launched (Swanson, 1984; Cernea, et al., 1985). The initiatives of both research and extension focus on farm management and the factors affecting farmers' daily decisions and overall strategies.

It has been hypothesized that these approaches can break down the traditional barriers between research and extension. On-farm research teams should themselves become the critical link: "Farm-management oriented research/extension personnel can serve in a *research and extension* capacity to work with *farmers and research scientists* in technology development" (Andrew and McDermott, 1985; italics mine).

The achievement of this admirable goal is a major challenge for the managers of both research and extension institutions. On-farm research cannot in itself solve the problems of technology transfer, or substitute for an effective extension system. Indeed, moving researchers off the station into the "space" conventionally occupied by extension and development institutions requires the careful rethinking of mutual roles and functions, as well as the development of new ways of working together.

This process often brings other organizational and managerial problems into relief. If either the research or the extension institution suffers from poor leadership, inadequate funding or poor staff morale, linking them will not solve the problem. If effective mechanisms for the joint planning and implementation of tasks related to common goals are not developed, information on farmers' needs will not be used effectively, no matter how many surveys, experiments, trials or demonstrations are carried out (Stoop, 1988). If farmers do not participate fully, the technology developed is unlikely to meet their needs.

Scope of This Analysis

This paper forms part of two studies being undertaken by the International Service for National Agricultural Research (ISNAR): the study of the links between research and technology transfer; and the research project on the organization and management of on-farm client-oriented research (OFCOR) in national agricultural research systems.

OFCOR is designed to establish closer links between research and resource-poor farm households. Numerous approaches to this type of research have been developed, including "cropping systems research", "farming systems research", "on-farm adaptive research", "farmer-back-to-farmer", "farmer-first-farmer-last" (Byerlee et al., 1988; Collinson, 1987; Gilbert et al., 1980; Harwood, 1979; Rhoades and Booth, 1982; Zandstra et al., 1981). What all these approaches have in common is a focus on farmers as the clients of research, an emphasis on diagnosing constraints and setting research priorities in the context of the whole farm system, the design of technological solutions in response to opportunities or constraints identified on farm, and the involvement of farmers at various stages in the research process.

The analysis is built around case studies of national agricultural research systems which have formally included OFCOR as a major activity and have at least 5 years' experience with this research approach (Avila et al., 1989; Budianto et al., 1989; Cuellar, 1989; Faye and Bingen, 1989; Jabbar and Zainul Abedin, 1989; Kayastha et al., 1989; Kean and Singogo, 1988; Ruano and Fumagalli, 1988; Solíz et al., 1989). Nine countries were included in the study: Bangladesh, Ecuador, Guatemala, Indonesia, Nepal, Panama, Senegal, Zambia and Zimbabwe. Improving cooperation between researchers, extensionists, development agencies and farmers was an explicit goal of most of the programs reviewed. A variety of mechanisms had been developed to link researchers and extensionists in the planning and implementation of various tasks. Nevertheless, forging effective, sustainable links across institutional barriers proved a major challenge.

The case studies review the experience of nearly 20 on-farm programs, organized in a variety of ways, following different approaches and using different methods. The word "program" is used loosely, to describe any organized on-farm activity; it does not necessarily imply the existence of a formal program analogous to a semi-autonomous, multidisciplinary commodity program of the kind commonly found in national and international institutes. The role of on-farm research as a means of strengthening links between research and extension was a key area of analysis in the case studies. It should be noted, however, that the studies were written from the perspective of research, and do not provide a detailed analysis of the extension institutions with which the on-farm research programs interact.

In the first section of this paper, the relationship between on-farm research and extension is contrasted in three countries — Guatemala, Nepal and Zambia. The second section draws on evidence from all nine countries to analyze the experience with six mechanisms for linking on-farm research and extension. The final section points out the lessons that emerge from the case studies for research managers using on-farm research as a means of strengthening the links between research and extension.

Assessing the Effectiveness of Linkage Mechanisms

The effectiveness of mechanisms linking on-farm research with extension will be assessed in terms of these questions:

1. How well does the mechanism, or group of mechanisms, facilitate the flow of information on farmers' conditions and needs to researchers — does it improve the system's responsiveness to the needs of its targeted clients?
2. How well does the mechanism facilitate the flow of information and techniques from the research system to resource-poor farmers — does it improve the system's capacity to transfer relevant technology?
3. How sustainable is the mechanism, given the various institutions involved?

Responsiveness to the needs of targeted clients. The diagnosis of farmers' conditions and needs is the basis for setting priorities and planning research. Informal and formal surveys, on-farm trials, meetings, field days and other special events all provide opportunities for researchers to learn from farmers (Biggs, 1989; Ewell, 1988). A number of mechanisms have been used to analyze farmers' needs and then carry the lessons learned into the process of planning and programming research on experiment stations (Merrill-Sands and McAllister, 1988).

Most approaches to on-farm research assign primary responsibility for the functions of diagnosis and feedback to social scientists from the research institution (Byerlee et al., 1980; Byerlee and Tripp, 1988; Zandstra et al., 1981). Some authors have envisioned a much broader role for extensionists in on-farm research programs as the principal "voice" of the farmer (Johnson and Claar, 1986; Johnson and Kellogg, 1984; Lionberger, 1986). Nevertheless, on-farm research programs have taken over this function in most cases precisely because the professional capacity of extensionists has been judged unequal to the task.

The case studies show that extension agents have participated in the processes of characterization and diagnosis of local farming systems primarily as informants. They have provided information on the agro-ecological conditions and farming systems in their areas as a preliminary basis for planning research; they have helped to locate farmers for surveys, experiments and field days; and in some cases they have served as enumerators. They have been seen as a resource — as a broadly distributed network of people in day-to-day contact with farmers. However, they have seldom been treated as equal partners, or given co-responsibility for setting priorities or channeling more detailed information into the research system.

Capacity to transfer relevant technology. In the countries studied, the tasks involved in adapting and transferring improved technology to farmers had traditionally been assigned to extension institutions. By developing on-farm research programs, the research institutions have taken on new responsibilities for working directly with farmers. This has changed the demands placed upon extension services: instead of demonstrating a uniform package of technology, extensionists are now expected to adjust the flexible recommendations resulting from on-farm research to suit local variations in agro-ecological and socio-economic conditions. This requires training and other resources which have often been beyond the capacity of the extension departments.

Some on-farm research programs have relied on the demonstration effect of on-farm trials and on informal communication among farmers to diffuse technology, with very little contribution from the official extension service. Others have used conventional mechanisms such as technical bulletins and field days to communicate the results of on-farm research to extension agents, who are then expected to diffuse them more widely; special projects have occasionally been set up to link on-farm research with the Training and Visit (T and V) system of extension. Still others have sought more direct collaboration and have defined explicit roles for both researchers and extensionists at established stages in an integrated approach to technology generation and transfer. The rationale for this integrated approach has been that if extensionists are involved in, or at least informed about, the on-farm research program, they will be much more knowledgeable about and confident in the technologies and recommendations produced and, thus, more committed to their transfer and diffusion.

Institutional sustainability. The case studies report several examples of links between research and extension that have not lasted. Many on-farm research programs have been developed with the support of international agricultural research centers (IARCs) and donors. Linkage mechanisms that have seemed very promising in a pilot project supported with special funding and expatriate staff have not always been successfully incorporated into the procedures of the institutions responsible for maintaining them after support has been withdrawn.

The most successful cases of institutionalization are those where links have been forged simultaneously at several levels of the administrative hierarchies of the organizations involved. Good cooperation at the field level is impossible to sustain unless regular opportunities to meet and work together are actively supported by management. Again, joint goals agreed upon by high-level coordinating committees cannot be realized unless specific operational procedures are worked out at both regional and local levels.

Three Case Studies

Out of the nine case studies, three exemplified markedly different degrees of integration between on-farm research and extension. Two cases, Guatemala and Zambia, lay at opposite extremes, while one, Nepal, was "intermediate", representing the kind of situation commonly found in developing countries.

Guatemala provides an example of an on-farm research program developed separately from the extension service, on the assumption that new technology adapted to farmers' conditions would diffuse spontaneously. The limitations of this approach led to the organization of a large project to bring extension into the process, over 10 years after on-farm research had been started. In Nepal, extension agents were involved in various on-farm research activities under the auspices of different agencies, but solid links had proved elusive. Heads of stations or programs had set up links on an ad hoc basis, but a high-level policy commitment and strong leadership from an integrated senior managerial group were lacking. The new national on-farm research program in Zambia was organized from the start with strong research-extension links at various levels of the administrative hierarchy, including the highest level. It is too soon yet to tell whether the

Zambian model is successful, but good progress has been made in integrating the research and extension systems, such integration being one of the hallmarks of successful agricultural technology generation and transfer in developed countries.

The material in this chapter draws extensively from Ruano and Fumagalli (1988) in the case of Guatemala; Kayastha, Rood and Mathema (1989) in the case of Nepal; and Singogo (1987) and Kean and Singogo (1988) in the case of Zambia.

Guatemala

Guatemala's national agricultural research system was totally reorganized in the early 1970s because the existing system for the generation and transfer of technology was not meeting the needs of an important group of clients. The agricultural sector of the Guatemalan economy is highly polarized: large-scale farmers, who constituted less than 1% of the population in 1970, controlled over 80% of the country's cultivable land. Most of their farms are located on good soils on the coastal plain or at mid-elevations, and specialize in the production of high-value export crops. This group has long had privileged access to modern technology, credit and inputs from public and private institutions.

The majority of rural households are concentrated in the highlands. Working on small plots, these small-scale farmers produce food crops both for home consumption and for sale. In the early 1970s, the capacity of this peasant sector to meet the demand for food in the rapidly growing urban areas was deteriorating, while imports were increasing.

Since the 1940s, research and extension services within the Ministry of Agriculture had followed procedures based on models from the USA (Mosher, 1957). Researchers developed programs within their disciplines according to their own interests and judgement. Extension was seen as a "top-down" program of adult education, spreading information about modern methods of farming. Neither was based on any analysis of the needs of particular groups of farmers. Some of the results were useful to large-scale farmers, but peasant producers received very little benefit.

ICTA: An institution integrating on-station and on-farm research. As one response to the mounting crisis in food production, the Instituto de Ciencia y Tecnologia Agricolas (ICTA) was founded in 1973 as a semi-autonomous research institute to generate, adapt and transfer technology appropriate to the conditions of small- and medium-scale farmers. A team of senior national scientists developed an integrated research system which linked on-station and on-farm research in a single process based on the diagnosis of farmers' conditions and needs. They drew heavily on the experience of the Office of Special Services (OSS), which had included extensive on-farm testing in the successful development of improved wheat varieties in both Mexico and Guatemala during the 1950s.

The pioneering institutional arrangements and working methods developed at ICTA had a major impact on on-farm client-oriented research in many other countries. However, no explicit, formal role was provided for extension in the initial plan.

Technology development system. Figure 1 illustrates how ICTA's system has structured both the flow of information from farmers into the research process, and the adaptation and transfer of relevant technology. The agenda for applied research developed by scientists in the commodity programs at the regional experiment stations is based on three types of input. The first is the basic and strategic research which is carried out by IARCs and universities, and the contributions of other public institutions and the private sector.

The second is an evaluation of farmers' needs through studies organized by ICTA's Socio-economics Department: both scientists and senior administrators participate in informal interdisciplinary surveys called *sondeos;* more detailed data on costs and returns are then collected from the farm records of a smaller sample of farmers. The third type of input is feedback

Figure 1. Diagram of the flow of information through ICTA's system for the generation, testing and transfer of technology

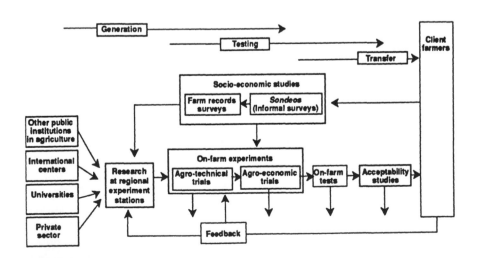

Source: Based on Ruano and Fumagalli (1988)

from on-farm research: all the technology produced by the stations is run through a standard sequence of on-farm trials, which are the responsibility of the Technology Testing Department. This department is organized as subregional teams, each consisting of five or six agronomists assisted by locally hired technicians.

First in the sequence of on-farm trials are multi-factorial experiments called "agro-technical trials". These are designed and implemented by researchers; the farmers contribute land, some labor and their assessment of the results. Next, the costs and returns of the most promising technologies are compared with those obtained using farmers' current practices in simpler experiments known as "agro-economic trials". Technology which passes this stage is then validated in on-farm tests, which follow a simple, standard design, on a larger number of farms. In theory, the information from all three stages is fed back into the process of planning and priority setting at the stations. Finally, surveys known as "acceptability studies" are carried out to see how many of the farmers who participated in on-farm experiments have actually adopted at least some elements of the new technology, and if so, which.

Here the process stops. The only systematic mechanism for transferring technology to the wider target group of farmers are field days for the neighbors of the participants and for extension workers. The assumption is that good technology spreads spontaneously through informal networks of farmers.

Does good technology spread spontaneously? There is evidence that some new crop varieties have indeed spread spontaneously. Over time, suitable inputs and management practices to accompany the new varieties have also been adapted to local conditions by farmers. An evaluation carried out in the La Maquina area of the coastal plain where an ICTA team had introduced an open-pollinated maize variety found that the extension service had played almost no part in its diffusion. The principal mechanism had been the on-farm tests, the results of which had been diffused through the neighbors and friends of collaborators. A second and much more modest influence was exercised by the commercial agrochemical companies promoting improved seeds and pesticides.

This kind of impact is concentrated in areas with high yield potential. Although the beneficiaries are small-scale farmers relative to the large-scale export-oriented sector, they are nonetheless those with relatively privileged access to resources within their communities. The much larger number of resource-poor farmers working on steep slopes and under other marginal conditions are largely left out.

Weak links with extension. These have been a major constraint on the adoption of ICTA's technology. The Dirección General de Servicios Agricolas (DIGESA), the national extension service for crops, did not change its philosophy and methods in response to the development of ICTA. The service lost its best professionals when ICTA was formed, and lost status relative to the new, highly visible, internationally connected organization (Gostyla and Whyte, 1980). Extension agents were less well educated than researchers, with significantly lower pay and status. They were not formally included in *sondeos*, planning meetings or other mechanisms built into ICTA's system for obtaining feedback from farmers. Researchers were responsible for all on-farm experiments.

DIGESA continued to follow the model of an adult education program, promoting modern methods to"ignorant" farmers (Ruano, n.d.). Local extension agents remained responsible for a number of bureaucratic tasks besides those directly connected with agricultural production. Until 1982, they played a key role in the processing and approval of applications for credit. In most areas, extension worked quite independently of ICTA. A survey conducted in 1982 found that very few extension agents could explain how the research system worked. Most did not know what its technical recommendations were or, if they did, could not explain their potential economic benefits to farmers (McDermott and Bathrick, 1982).

Good informal working relations developed between ICTA's technology testing teams and local extension agents in some areas, particularly where technology in high demand from farmers was becoming available (Whyte, 1983). ICTA personnel depended on extensionists for the selection of collaborators, and to widen their area of influence. DIGESA and ICTA have different approaches to field days, but often combine them in practice.

Formal mechanisms to involve extension agents in the research work proved difficult to sustain. In one region, extension agents were invited to take part in the implementation of on-farm tests. For 2 years, a special course for researchers new to ICTA — the Curso de Producción Agricola (CAPA) — included a subcourse for extension agents, so that they would understand the three stages of ICTA's work and convey its results to farmers. Both of these initiatives fell foul of the same institutional bottleneck. Although they were based on formal agreements between the directors of ICTA and DIGESA, local extension managers did not reduce the load of other tasks agents were expected to perform. Work with ICTA came to be seen as an extra burden which could not be sustained.

A new joint program. To improve matters, the Proyecto de Generación y Transferencia de Tecnologia Agropecuaria y Producción de Semillas (PROGETTAPS), a new program for the generation and transfer of technology and the production of seed, was established in 1986. The program is based on the concept of close links between research and extension (Ortíz, 1987). Funded by the International Fund for Agricultural Development (IFAD) and the Inter-American Development Bank (IDB), it draws on ICTA's earlier experience in collaborating with a World Neighbors project in San Martin Jilotepeque (Ruano, pers. comm.). This program was designed to be implemented jointly by ICTA, DIGESA and the Dirección General de Servicios Pecuarios (DIGESEPE), the national livestock extension service.

Local extension workers are given the title "promoters", responsible for promoting specific technologies, not for providing general technical assistance. In contrast to the situation in many other countries, the task of promotion has become the full-time responsibility of the extension agents working in the program. Selected farmers, known as "rural leaders", are trained in the management of new technologies and hired on a half-time basis. One, two or three promoters are tied to each research scientist on the technology testing teams. Each of these promoters is expected to work with 15 to 20 rural leaders. Technology which has already been validated in on-farm trials is demonstrated in "transfer parcels" managed by the rural leaders on their land. Each rural leader then supervises similar demonstrations on the farms of 20 neighbors. Through this "branching tree" approach, the work of each on-farm research scientist is expected to reach up to 600 farmers. Farm records surveys permit researchers to evaluate the economic benefits of a new technology, to monitor its adoption and to provide guidelines for credit.

The project has set up several support activities, including seminars and workshops for training the promoters. Funding for new staff, vehicles and other facilities is provided. A national coordination committee and regional subcommittees have been set up by government decree.

In short, the program is an attempt to draw extension into a structure and approach based closely on what ICTA has already developed. It has expanded the network of farmers exposed to new technology through on-farm trials, but does not envision qualitatively different extension methods. In other words, it is an attempt to broaden and institutionalize the concept of OFCOR as the basis for the diffusion of technology.

Early reports on the implementation of PROGETTAPS indicate that good progress has been made. In only 2 years, research teams have carried out validation trials at 3000 sites and rural leaders have laid out about 8000 transfer parcels. The program appears to be reaching the very poor, and farmers' demand for new technology is such that the program has had to organize small-scale seed production units (Ruano, pers. comm.). However, success in the longer term, particularly if external funding is reduced, will depend on close collaboration between institutions with a disappointing history of cooperation. They will need to institutionalize common objectives, a uniform operational approach, and integrated work plans.

Conclusions. ICTA's past provides a clear example of an innovative on-farm research program whose success in meeting the broad range of needs of its target clients was limited by poor links with extension. Its present demonstrates an imaginative approach to overcoming linkage problems.

Responsiveness to the needs of targeted clients. Until PROGETTAPS was formed, the extension service played a very limited role in diagnosis and feedback. The primary responsibility for bringing information on farmers' conditions and priorities into the research processes was given to ICTA's Socio-economics Department. The social scientists developed an innovative approach, but the department lost most of its senior scientists in the early 1980s after the departure of its first expatriate director. For several years, ICTA was unable to channel a continuous flow of information on changing rural conditions into its research program; nor was extension offered any role in this process.

The agronomists in the technology testing teams were in constant communication with the farmers with whom they worked, and had informal contacts with local extension agents. They provided feedback into the rest of the research system on the performance of particular technologies under farmers' conditions.

Capacity to transfer relevant technology. ICTA has successfully transferred new technology — primarily new crop varieties — onto the farms of producers of basic food crops who had not previously benefited from public sector research. In general, the beneficiaries are the less disadvantaged members of their communities, with privileged access to resources. This subgroup of clients is in a good position to pick up new technology through informal networks and to purchase the necessary inputs. The lack of an effective extension system has limited diffusion to farmers with more limited resources working in more marginal areas. The new joint PROGETTAPS program promises to broaden the coverage and increase the efficiency of the same basic system of diffusion by demonstration.

Institutional sustainability. Before 1986, attempts to link ICTA's on-farm research with extension proved unsustainable. The PROGETTAPS program has been initiated with substantial external financing. Its long-term effectiveness will depend on how solidly it can be incorporated into the regular procedures and budgets of the three institutions involved.

Nepal

Agricultural development in Nepal faces severe constraints. The mountainous topography and lack of roads inhibit communications and make inputs expensive and difficult to obtain. Despite significant investments in research since the early 1970s, production of basic food grains between 1970 and 1981 increased at an average annual rate of under 1% — far below the annual population growth rate of 2.7% in the same period (Yadev, 1987).

One of the few areas with a relatively high potential is the Tarai, the lowland plain along the border with India. The control of malaria and the building of roads and other infrastructure opened up the Tarai for settlement, starting in the 1950s. Researchers concentrated on developing high-yielding crop varieties appropriate for conditions in the Tarai. However, most of the rural population lives and farms in the lower potential, heterogeneous hill districts. The generation and transfer of technology capable of increasing their output and incomes has presented major difficulties for on-farm research and extension programs.

-**The institutional structure.** In Nepal, extension and most research is carried out under the Ministry of Agriculture. On-farm client-oriented research is organized in several different research departments and organizations. In each case, extensionists have been asked to perform a role.

The basic units for the extension service are the 75 political districts into which the country is divided. Agricultural District Officers direct extension workers outposted in rural areas, each comprising several villages. Each extension worker is theoretically responsible for an average of 2500 farm households. Technical programs are planned and coordinated both at the national level and in the five development regions into which the country is also divided.

Village-level extension is not an established professional career. At one time, all students of agriculture were required to serve as extension agents for 2 years—a year as an assistant in the middle of their secondary training, and another year as a technician —before being admitted to the university. Then, because of manpower shortages, permanent extension positions were created, with no chance of advancement into research. Salaries and benefits for these extension workers are low, and staff turnover is high. Only during the 1980s have professional subject-matter specialists been appointed in some districts as a part of the T and V system.

The national agricultural research system in Nepal is organized as departments and commodity programs, supported by a network of experiment stations and farms. The case study highlights the on-farm programs of the National Rice Improvement Program, the Cropping Systems Program — which subsequently became the Farming Systems Research and Development Division — and the externally funded Lumle and Pakhribas Agricultural Centres.

Each program has developed its own on-farm and outreach agenda. Each has different types of links with extension, but all participate in nationally organized on-farm demonstration programs as well. Research and extension are coordinated informally at both national and regional levels, but no formal mechanisms for joint planning or evaluation across the sector had been developed at the time of the study.

Outreach activities of the National Rice Improvement Program. All the major commodity programs in Nepal have outreach activities in the areas immediately surrounding their principal experiment stations. These activities were organized in response to requests for greater technical support from the regional extension officers. The outreach activities also provide the scientists based at the stations with opportunities to obtain first-hand experience of the issues being faced by farmers and extension agents in the field.

The outreach activities of the National Rice Improvement Program, highlighted in the case study, are typical of those which are found in the commodity programs. They are part of a larger World Bank program centered on the establishment of a T and V system of extension. Two outreach officers located on the research station administer an on-farm research program designed to adapt crop varieties and other technology to local needs. The implementation of the trials is delegated to local extension agents. The outreach officers provide back-up to the rice subject-matter specialists of the extension service, who in turn provide technical support to the extension agents at village level. They have also organized regular bi-monthly and bi-annual meetings at the station, when research specialists and the senior extension staff discuss problems identified in the field, potential solutions, and plans for future research. In addition, village-level extension workers are brought to the station for training in problem identification, methods for on-farm trials, and the background of new recommendations.

The Cropping Systems Program. The Cropping Systems Program operated from 1977 to 1985 within the Agronomy Division of the ministry. The program's approach was developed by the International Rice Research Institute (IRRI), working in cooperation with national agricultural research systems in a network covering Asia (de Datta et al., 1978). An integrated approach to research and extension was designed, with the aim of replacing farmers' current production practices with improved cropping systems over large areas. Research and extension were given precise, predetermined roles.

Rigid approach through production programs. Those who developed the approach started from the assumption that a "submissive approach", which depended entirely on improved technology lying within farmers' existing limitations, would be unlikely to have significant effects on food production. Instead, they proposed an "interventionist approach", combining improved technology with packages of credit, inputs, irrigation and other improvements. As these services were supplied by separate government organizations in most countries covered by the network, special production programs were set up as "buffer" institutions to concentrate the necessary resources and to coordinate their use (Zandstra et al., 1981).

The researchers developed a uniform set of methods for the development, testing and promotion of new technologies based on improved grain crop varieties. Target areas for development were selected before research sites, the whole process being directed towards specific, production-oriented goals (Denning, 1985). The technology generation and transfer process followed a set sequence of steps:

1. site selection
2. site description—benchmark surveys, crop-cut studies, farmer interviews, and farm management studies
3. design of improved cropping systems, under controlled conditions
4. cropping systems testing, in farmers' fields
5. pre-production evaluation — multi-locational on-farm testing of promising technical alternatives, implemented in cooperation with extension
6. production programs, to diffuse the innovations over large areas, under the management of extension and development agencies

Extension became involved only at steps 5 and 6. Pre-production verification trials were designed by researchers, but extensionists were usually involved in their implementation. Eventually, responsibility for managing the programs was transferred to extension and development agencies.

Success in the Tarai. The program in the Tarai was designed to promote packages of technology based on improved rice and other crop varieties, most of which were already available "on the shelf". National scientists and their expatriate advisors determined which varieties could be fed into on-farm research. Interdisciplinary teams of researchers were given responsibility for the early stages of the process—the selection of sites, the diagnosis of local conditions, the design of improved cropping systems, and the preliminary testing of these systems in farmers' fields.

Detailed manuals explained how extension personnel were expected to carry out their part of the process — the broad testing of promising technology and the administration of input supplies and credit. Senior extension officers were represented at the planning and review meetings held before each cropping season. Initially, the researchers ran pre-production verification trials in pilot areas. Little by little, procedures were simplified and responsibility was handed over to local extension staff.

The researchers were concerned to maintain the consistency of the data collected, and thus discouraged adaptation of the content or design of trials to local circumstances. Analysis was handled centrally, with the result that extensionists could not easily use the results of the experiments they had implemented. Modifications to the original packages were made, but on the recommendations of the researchers, not the extensionists.

The highly structured Cropping Systems Program was reasonably successful in the two of its five sites which were located in the relatively high-
potential Tarai. The responsibility for production programs in 22 districts
was passed on to extension after 6 years.

Difficulties in the hills. Developing appropriate technology and forging
links with extension proved much more difficult in the heterogeneous,
densely populated hill regions, with their poor communications facilities.
Once again, the goal of the researchers was to have a dramatic impact on
production. Sites were chosen on the basis of rapid reconnaissance tours
using two criteria — high theoretical potential for the technology the
program intended to promote, and low current use of improved technology
of any kind. There was no size limit — that is, no specified number of
villages or households — and no relationship between the sites selected and
the operational zones of extension.

This program encountered several implementation problems. The extension workers regarded the trials they were expected to administer as a
burden on top of their regular work. Researchers complained that prescribed steps in the methods to be followed had been omitted, and that extensionists had been careless with trial management and data (Lipinski and
Rizal, n.d.). The basic problem was that extension had been handed an
impossible task. The high-input technology that researchers were promoting could not realistically be supported through production programs
under resource-poor conditions.

Reorganization: Farming Systems Research and Development Division.
The advisors and planners concluded that the technology being promoted
in the hills was too narrowly based on the major grain crops. A broader
range of more flexible technologies was needed to provide farmers with
productive alternatives. In 1985, the Cropping Systems Program was reorganized with a broader farming systems mandate and elevated to the status
of a fully fledged research division. Known as the Farming Systems Research and Development Division, it works exclusively in the hills.

The Cropping Systems Program had been a special program with a production-oriented mandate. Researchers from various commodity programs
had worked closely with local extensionists in the target areas. The creation
of the new independent division weakened the links with other research
divisions and with extension. Its field assistants have been employed
directly and, so far, have had almost no contact with the Agricultural
District Officer or other extension personnel in the districts where they
work.

Socio-economic research, which had been an integral part of the Cropping Systems Program, was recently separated from the Farming Systems
Division to form the Socio-economic Research and Extension Division. In

spite of its name, this group has not worked with extension, except for a single survey of the methods used by different agencies. Its professionals feel over-extended, and have lobbied to have their mandate narrowed by dropping the word "extension" from the name of the division.

Lumle and Pakhribas Agricultural Centres. The Lumle and Pakhribas Agricultural Centres were established in 1968 and 1973 respectively, in different areas of the hills. Funded entirely by the British Government, their initial purpose was to support the resettlement as farmers of Gurkha mercenaries returning from the British army. The centers developed their own extension activities to serve specific target areas, and organized both on-station and on-farm research.

The centers later expanded their mandates to include the provision of technology to all farmers, covering larger areas. They have taken advantage of the flexibility provided by external funding to develop some innovative methods and procedures, including the involvement of extension in the planning and implementation of research. Nevertheless, neither of them has established close working relationships with the ministry's regular extension staff, although both centers have recently been officially integrated into the public sector national agricultural research system.

The Lumle Centre has concentrated its work in a single target area surrounding the station. Originally, each commodity section at the center organized its own extension efforts. A farming systems research section was set up with its own field staff, which implemented on-farm trials in selected subdistricts. As the center's activities multiplied, farmers became unsure whom to ask for information on specific topics. In response, the center created a separate extension section responsible for synthesizing information from the researchers and passing it on to farmers. This service completely replaced the work of the ministry's extension agents in the target area. Links with extension in the larger region were developed only in the mid-1980s, with the naming of outreach research staff to feed technology into a T and V program.

The Pakhribas Centre has its own extension programs in two separate target areas, serving a total of about 9000 households by 1986. The center has also established on-farm research as a mechanism for feeding information to extension in the four districts covered by the Koshi Hills Area Rural Development Project (KHARDEP).

Both centers have set timetables for integrating their work more closely with that of the ministry, including extension.

National on-farm demonstrations. Two different types of on-farm trials — farmer field trials and minikits — are routinely implemented through the ministry's extension department. Farmer field trials are standardized tests of promising technology. They are designed by scientists in the commodity programs, and run either by researchers on regional stations and farms, or by extensionists on farmers' fields. Data are collected, sent back to the commodity programs and analyzed centrally. For the local extension agents, the trials are simply one more routine task. They have not been authorized to modify the designs in any way, and the results are never analyzed in terms of local conditions. The usefulness of the trials at national level has also been limited. The trials clearly show a wide gap between yields on stations and on farms, but they do not provide enough information on farm-level conditions to identify specific constraints or suggest potential solutions.

Minikits were initially designed as a relatively cheap and easy way to provide feedback to the breeding programs on the performance of different varieties and advanced lines under farmers' conditions. Small packets of seed, sometimes accompanied by measured amounts of fertilizers or pesticides, are distributed through extension to farmers, along with a form which the farmer is expected to fill out with his or her reactions and return by mail. In most parts of the country, few cards are returned and little or no analysis is done of the data from those that are. Extensionists have a role in administering the program, but are not given enough discretion to provide useful feedback. The minikits are an effective mechanism for the wide distribution of new seed, but they are ineffective as a research tool and as a means of demonstrating new technology for extension purposes.

The integrated research and extension programs at the Lumle and Pakhribas Centres use minikits, in a modified procedure, as a tool within their target areas. Instead of distributing just one kind of improved seed, they include local varieties in the package. The extension agents follow up with the farmers and collect the forms, which are analyzed at the local station before being sent on to the national program. Feedback is effectively stimulated on several levels.

Group treks. Systematic feedback from farmers is difficult to obtain in Nepal, given the difficulties of communications and travel. Several on-farm research programs have met this challenge by organizing group treks at regular intervals. Senior scientists and on-farm researchers travel together through the target areas, interviewing farmers and officials. They assess local conditions and constraints, and put together work plans for on-farm research on the spot. The Lumle and Pakhribas Centres, where this approach was first developed, include senior extension staff on their treks. Managers of the Farming Systems Research and Development Division sometimes invite Agricultural District Officers on their treks as a formal courtesy, but have on the whole made much less effort than have other programs to draw on the experience of extension personnel.

Conclusions. Personnel from extension participate in on-farm research in Nepal in various ways, but formal links have proved difficult to institutionalize. Although research and extension operate within the same ministry, links at national level are weak. Apex management has not played a strong role in encouraging the integration of the research and technology transfer system as a whole. A recent reorganization which has strengthened the independence of the research branch has, if anything, reduced the formal opportunities for joint planning and coordination.

Responsiveness to the needs of targeted clients. All the on-farm research programs examined in Nepal had accepted primary responsibility for diagnosing needs and constraints at the farm level. The group trek is the primary mechanism for bringing senior researchers directly into contact with farmers on a regular basis. Extension has played only a supporting role.

Capacity to transfer relevant technology. Outreach programs have provided a means of getting information from research into the hands of extensionists, both through the T and V system and through KHARDEP. The support of extensionists has been enlisted to extend the coverage of on-farm research. Farmer field trials and minikits have brought new varieties and other technology to the attention of large numbers of farmers. However, these mechanisms have not been flexible enough to give extension an active role in adapting technology to local conditions.

Institutional sustainability. Agricultural research in Nepal, and on-farm research in particular, has been heavily supported by donors and IARCs. Specific linkage mechanisms, such as the group treks at the Lumle and Pakhribas Centres and the production programs of the Cropping Systems Program, have been dependent on external funds. These mechanisms have proved difficult to institutionalize in the ministry, with its highly restricted budget for operations.

Zambia

In Zambia, research and extension are the two branches of a single administrative structure within the Ministry of Agriculture. On-farm client-oriented research has been introduced as a national program in the research branch. Field work is organized through semi-autonomous provincial operational units known as Adaptive Research Planning Teams (ARPTs). Each team carries out on-farm research in a number of small areas which are selected to represent agro-ecological "recommendation domains". The work of these teams is intended to complement that of the Commodity and Specialist Research Teams (CSRTs) which are responsible for applied research on the experiment stations. The managers of the ARPT program have placed a great deal of emphasis on institutional issues and, of the nine countries studied, Zambia has developed the most elaborate set of mechanisms to link research and extension.

The extension service in Zambia is based on the T and V system and is administered by Agricultural Officers at provincial and district levels. Although a formal structure has been created to support this extension system, in many parts of the country its implementation has been inhibited by low population densities and organizational problems.

Improved links with extension: An explicit goal of ARPTs. Before ARPTs were set up, farmers' needs were brought to the attention of station-based researchers through provincial research tours, followed by meetings of the Provincial Experimental Committees. These tours enabled junior and senior extension staff to meet researchers, but they were not systematic or frequent enough to provide accurate information for setting research priorities. At the meetings, more time was spent discussing administrative problems and bottlenecks than technical research issues.

On a practical, day-to-day level, there was little interaction between research and extension. Extensionists saw the work being done on experiment stations as irrelevant to the needs of the farmers with whom they worked; researchers blamed extension for not transferring technology to farmers. When the ARPTs were set up in 1980, two explicit goals of the program were:

1. To draw the extension staff into the process of generating and adapting technology
2. To pass information on to extension, credit and marketing institutions

Each provincial ARPT is funded by a different donor, and has experimented with different methods and procedures for organizing on-farm research and linking with extension. The ARPT program was intended to support extension workers in various ways, particularly by sharpening the focus on the conditions of small-scale farmers and the logic of their decision making. Much has been learned, although surveys of extensionists have revealed widespread confusion as to whether on-farm trials are an adaptive phase of research or a demonstration phase of technology transfer.

Complementary links between research and extension have been established at various levels of the administrative hierarchy. The major points of contact are summarized in Table 1.

National policy and coordination. Cooperation between on-farm research and extension has received high-level support within the Department of Agriculture. Senior staff, including the Assistant Director of Agriculture for Extension, were directly involved in setting up the ARPT program. The Assistant Directors of Research and Extension have adjacent offices. For

Table 1. Zambia: Links between on-farm research and extension at various levels of the administrative hierarchy

Administrative level	Linkage mechanism
National administration	The Assistant Directors of Agriculture for Research and Extension have been involved in the on-farm research program since it was first established, and confer frequently.
Provincial administration	Provincial ARPT committees are chaired by the Provincial Agricultural Officers, who are the heads of extension in each province. Meetings are attended by subject-matter specialists from extension. The committees recommend sites for on-farm research and review the on-farm client-oriented research programs. The committees have not been as effective as their creators hoped.
ARPT provincial teams	A Research-Extension Liaison Officer is assigned to each provincial team. A professional employed by extension, he or she is responsible for facilitating the flow of information in both directions.
On-farm research teams	The Trials Assistants, who implement surveys and on-farm experiments, are seconded to ARPTs from extension.
Local extension workers	Contacts between researchers and local extension workers outside the research areas have been limited.

several years while the ARPT program was first being developed, its national coordinator had his office in the same building as well. This close contact between policy makers and senior administrators permitted frequent consultations over problems as they arose.

Coordination at provincial level. Provincial ARPT committees were set up as a forum for the joint planning and review of on-farm research and extension at the operational level. The meetings are chaired by the Provincial Agricultural Officer, who is the key figure responsible for the ministry's activities in each province, and are attended by both researchers and subject-matter specialists from extension. In theory, these committees are a critical linkage mechanism, but in practice their record has been disappointing. The only kind of decision on which they have had much impact has been the selection of target areas for on-farm research. Reviews of the research programs have been perfunctory, and there is little evidence that plans have actually been altered in response to comments from extension staff. Nevertheless, the committees have kept the subject-matter specialists informed about the purpose and progress of on-farm research activities.

Role of Research-Extension Liaison Officers. In early discussions of the composition of ARPTs, it was suggested that senior professionals from extension should be included as fully fledged members. This suggestion was adopted, but there has been no universal agreement as to what the job description of these officers should be.

The first Research-Extension Liaison Officer, an expatriate, was appointed to the team in Central Province, with funding from the United States Agency for International Development (USAID). He thought that neither the leadership of the ARPT nor the FSR methodology developed by the Centro Internacional de Mejoramiento de Maiz y Trigo (CIMMYT) involved extension sufficiently, and he worked to broaden its role. He stressed the importance of taking technology through a testing stage in close cooperation with the local extension workers. He organized training workshops, demonstrations and field days, and also started a monthly newsletter for extensionists.

In other provinces, the dual responsibilities of the Research-Extension Liaison Officers led to delays in recruitment and confusion over the job description (Hudgens, 1986). For example, no liaison officer was appointed to the team in Eastern Province until 1986, partly because the Farm Management Officer of the World Bank's extension program had nearly identical terms of reference. In fact, however, the latter spent almost all his time organizing the T and V system. A long delay in appointing a liaison officer for Luapula Province hindered interactions with extension. Little by little, the Research-Extension Liaison Officers demonstrated their usefulness, and by 1986 six of them — foreigners as well as nationals — were on the ARPT staff. Interest in filling the posts increased as the provincial ARPTs acquired technologies that were ready for broader testing and validation.

Use of extension workers as Trials Assistants. The single most important linkage mechanism was developed on an ad hoc basis. The program organizers did not at first have a clear plan to post technicians to the research areas to supervise the day-to-day operations of on-farm research. They did not really face this issue until they began to plan the trials for their first major field season in 1981. Rather than hire technicians directly, they decided it would be cheaper and more effective to use extension personnel seconded on a full-time basis.

These people play a critical role in the on-farm research process. They are usually from the areas where they work, speak the local languages, understand local farming practices, and serve as an effective link with village communities. They are responsible for implementing on-farm trials, and also assist in the organization of field days to diffuse the results.

Some extensionists without diplomas are recruited, but the standard of competence is generally high. Most of the Trials Assistants regard the opportunity to work in research as a privilege. Nevertheless, it has taken time to train them to become effective research technicians. When the ARPT program was beginning, training was conducted centrally, with a course for all Trials Assistants given at the central research station. Subsequently, the provincial teams assumed responsibility for providing informal training because it was thought that this would help develop stronger regional teams. The original idea was to rotate local extension workers through the ARPT program, to expose them to the research process and make them familiar with the new technology. In fact, the research teams try to retain them for as long as possible, to save the expense and trouble of constant retraining.

Trials Assistants are paid by the extension branch but supervised by researchers. This joint jurisdiction leads to some conflicts. For example, critical repairs to field housing were delayed while the two administrators argued over who should pay. Nevertheless, good communication at the provincial and national levels makes it possible to resolve issues of this kind before they become serious problems.

Links with non-ARPT extension workers. Contacts between researchers and local extension workers who do not work directly with an ARPT are limited (Edwards et al., 1988). Local agents are used as the main informants in informal, preliminary surveys carried out to demarcate farming systems and recommendation domains. They also help identify new research areas by introducing researchers to farmers and acting as interpreters. Once the research programs are established, however, even routine communications prove difficult to sustain. In Central Province, for example, only half of the extension staff regularly received the newsletter produced by the Research-Extension Liaison Officer for their benefit. Informal contacts between Trials Assistants and their colleagues who are local extension workers have been useful, but this influence has not extended beyond the research areas.

Conclusions. The ARPT program in Zambia has made significant progress in forging links with extension at various levels from the field up to the top of the bureaucracy. However, even in this situation, where senior research managers have given priority to developing strong links through on-farm research, there are still problems. The different methods employed in research and extension have led to problems of overlap and inadequate coordination. Various shortcomings have been identified for each of the linkage mechanisms, and important differences in attitudes and in organizational culture remain. The local extension workers are overworked and underpaid, and staff turnover is high.

Responsiveness to the needs of targeted clients. Extensionists at various levels have opportunities to bring farmers' perspectives and needs into the research process. The Trials Assistants are in constant contact with the farmers who cooperate in the on-farm research program. Nevertheless, as in many other cases reported from the nine countries, it has proved difficult to capture the results of this experience adequately. Only a few of the provincial ARPTs have systematically included the Trials Assistants in their annual research planning and review processes.

The primary responsibility for feedback lies with the social scientists in the ARPTs. The sociologists are organized as a special unit which conducts studies on a multi-provincial basis. They also provide support to the provincial teams for particular pieces of research. Economists are assigned directly to most teams, to conduct surveys and analyze the results of experiments; rather than use local extensionists, they hire and train their own enumerators. Some scientists argue that the economists on the teams should be replaced by Research-Extension Liaison Officers, who would be agronomists with some training in economic analysis.

Capacity to transfer relevant technology. In the early years, the ARPTs concentrated on the development of technology, on experiment stations and in farmers' fields. The program was only 6 years old at the time of the study, and the process of verifying promising results in broader on-farm tests was just starting. The choice of sites had been organized through local extension workers, under the coordination of the Research-Extension Liaison Officers. Where possible, demonstrations were located on the land of the contact farmers working with the T and V system. Lengthy discussions on the technology to be demonstrated were held with the subject-matter specialists. It was still too early to assess the effectiveness of the transfer process.

A variety of mechanisms are used to transfer preliminary information from the ARPTs to extension workers. Researchers participate in training courses for extensionists. ARPT agronomists and subject-matter specialists collaborate in the revision of formal recommendations. Scientists from both branches contribute material to newsletters for the field-level staff.

The T and V system creates incentives and formal settings for interaction, but also places very strict controls on the time and activities of extension workers. Unless they work directly in the research areas, they have few opportunities to receive information from ARPT researchers outside a few formal events.

Institutional sustainability. The ARPTs and extension depend on several donors with different approaches and priorities. Although formal linkage mechanisms have been put in place at national level — Provincial Coordinating Committees, Research-Extension Liaison Officers, the secondment of Trials Assistants to provincial ARPTs — their effectiveness has varied considerably among teams. Moreover, a great deal of administrative time has been spent on keeping critical linkages functioning. A strong commitment to research-extension links by senior administrators will be required if these are to be sustained once donor support ends.

Mechanisms Linking On-Farm Research and Extension

Six types of linkage mechanism were identified in the case study programs. They are not mutually exclusive and are usually found in various combinations with one another. These linkage mechanisms are:

- informal cooperation at field level
- national and regional research-extension coordinating committees
- participation of extension field staff in the implementation of surveys and trials
- participation of senior extension specialists as scientists in on-farm research, or of researchers as outreach officers in extension programs
- participation of on-farm research staff in rural development projects
- integrated on-farm research and extension programs

The first and second linkage mechanisms provide opportunities for members of staff to talk — to exchange information and ideas with each other and to plan joint activities. Such mechanisms are essential, but they must be backed up with more formal arrangements if shared programs are to be effective.

The third and fourth mechanisms involve the secondment of staff between extension and on-farm research programs. Direct collaboration of this kind is an effective way to pool experience and to get on-farm research activities moving. In the longer run, joint staffing often proves difficult to administer as seconded personnel lose their identity and become isolated from normal career opportunities in their parent institutions.

The last two mechanisms involve the joint participation of research and extension in integrated programs. This might seem to be the ideal solution, but in practice it is difficult to maintain the focus and continuity of research goals in the face of the strong, short-term pressures to produce quick results experienced in a development project.

In this section, the experience with these six mechanisms is discussed, and their effectiveness is assessed in terms of three basic criteria:

1. Their responsiveness to the needs of targeted clients
2. Their capacity to transfer relevant technology
3. Their institutional sustainability

Informal Cooperation at Field Level

Examples from the case studies. The on-farm research field staff in all the programs studied depended heavily on the informal cooperation of local extension agents for assistance in such areas as securing the cooperation of local leaders, identifying collaborators and organizing field days. Obviously, the success of any link depends on good working relationships between the people involved. Nevertheless, informal exchanges of information between people cannot by themselves serve as dependable linkage mechanisms.

As the experience in Guatemala demonstrates, informal cooperation must be supported by formal mechanisms, or researchers and extensionists will inevitably drift into the routine procedures of their parent institutions. In turn, many of the formal mechanisms function best when informal cooperation is already strong.

The Programa de Investigación en Producción (PIP), the on-farm research program in Ecuador, provides another good example of the limitations of unsupported informal cooperation. Several of the provincial PIP teams had shared offices with extension agents from the Ministry of Agriculture. They consulted each other about issues such as the selection of farmers, and organized joint field days, but various barriers prevented close collaboration.

First, the extension system was divided into operational regions which did not correspond with the recommendation domains developed by PIP. Second, the extensionists' experience in conventional programs had put them in contact with relatively large and prosperous farmers, not the resource-poor target group PIP was trying to reach. Third, the national extension program had been extensively reorganized several times; the resulting shifts in responsibilities made it difficult for researchers to develop and maintain working relationships with senior specialists. Finally, the day-to-day operating procedures of the two institutions did not mesh easily. The field extension workers were busy with their own tasks, and their budgets were limited. Their schedules did not give them enough flexibility to visit research sites with any frequency, even if the on-farm research teams offered transportation.

Assessment. The following assessment of informal cooperation at field level as an effective linkage mechanism is based on the three criteria listed above.

Responsiveness to the needs of targeted clients. Informal contacts with extension agents and other officials with experience at village level are a valuable first step through which on-farm researchers can learn about local farming systems and the constraints faced by farmers. They can also provide valuable introductions into the local community. Nevertheless, care must be taken to avoid introducing extensionists' biases into the research agenda. Extensionists often work with relatively prosperous farmers who are influential members of their communities. Over-reliance on their assistance can bias the samples and research priorities selected away from the needs of resource-poor farmers (Biggs, 1989; Ewell, 1988).

Capacity to transfer relevant technology. Informal field visits, which are supplemented with regular events such as field days, can be valuable mechanisms for transferring technology to extensionists in the immediate areas where on-farm trials are conducted. New crop varieties and some other technologies will then diffuse spontaneously through the informal networks of farmers. Nevertheless, as the experience in Guatemala shows, extension activities which are based on more formal links are necessary to transfer more complex technologies and to reach clients in marginal areas.

Institutional sustainability. Links which depend on informal, personal contacts between individuals fluctuate in their effectiveness not only according to changing circumstances in the field, such as staff turnover, but also according to the degree to which they are encouraged and supported at more senior levels. They are often invoked as evidence of a working relationship when in fact the institutions involved have not succeeded in developing more permanent mechanisms for cooperation.

National and Regional Research-Extension Coordinating Committees

Examples from the case studies. Coordinating committees with members from both on-farm research and extension institutions had been set up in several of the case study countries at both national and regional levels.

National coordination in Zimbabwe. Prior to independence in 1980, research and extension in Zimbabwe were organized to serve the needs of European farmers in the large-scale commercial sector. A major policy of the new government was to expand their mandates to meet the needs of African farmers in the communal areas. The communal areas are a legacy of colonial land policy, which authorized the private ownership of commercial farmland for the benefit of the white settlers, and recognized traditional communal patterns of land tenure for the African population in the remaining, more marginal areas of the country. Today, the communal areas consist of 170 separate territorial units. About 760 000 households farm and raise livestock on this land, much of which has very low productive capacity.

The Department of Agricultural, Technical and Extension Services (AGRITEX) was formed in 1981 by uniting the staff and facilities of two organizations. One of these organizations had served the commercial farmers and had long worked in close association with research, while the other had been a much less technically oriented division of the ministry responsible for Tribal Trust Lands, which had supported the African farmers.

There was a substantial exodus of experienced staff during the reorganization. Nevertheless, AGRITEX was one of the few agencies with an established structure in the communal areas, so heavy demands were placed on it by numerous agencies trying to comply with political directives to work there. Among the most demanding were the semi-autonomous institutes of the Department of Research and Specialist Services (DR and SS) of the Ministry of Lands, Agriculture and Rural Resettlement. These institutes had set up entirely separate and uncoordinated on-farm research programs.

Various seminars and workshops to address the problem were organized sporadically, but there was no forum for regular consultation or coordination until the Committee on On-Farm Research and Extension (COFRE) was established in 1986 at the initiative of research and extension staff working in the communal areas. The committee consists of the deputy Director of AGRITEX, and senior representatives from each of the research institutes of DR and SS working in the communal areas. It has been effective because its members have the authority to implement the decisions made. It has also been strongly supported by the Directors of DR and SS and AGRITEX, and resources have been allocated as required to carry out joint field activities.

The first coordinating body to cut across the decentralized structure of DR and SS, the committee immediately had a positive impact in several areas. It published a general directory of on-farm trials and demonstrations, to avoid overlap and duplication of effort. It organized joint field monitoring tours for senior staff from both research and extension. Specific research proposals and extension recommendations are now discussed at subcommittee meetings of specialists in the major commodities. This is a way of getting proposals screened and, if necessary, modified at an early stage, at a forum where it is not humiliating for a scientist to back down. Meetings between research and extension staff are held in each province to discuss their results and plans in the light of the comments made by the subcommittee. Workshops on special topics are held at intervals. The coordinating committee has been well received because it ties national plans to the concrete products of both research and extension at regional level.

Regional committees. In Zambia, among the mechanisms linking the ARPT on-farm research program with extension are the Provincial Coordinating Committees. As we have already seen, these have not been as effective as was first hoped. Few extension administrators or senior staff realize the power they could wield by taking a more active role in their meetings. Nevertheless, the committees have been far more effective than their counterparts in Guatemala and Ecuador, which are regional committees in form only. They have had no effective influence and seldom even meet.

Assessment. Coordinating committees can be an effective linkage mechanism if several conditions are met. At the very least, the objectives of the committee must be clear and there must be general agreement among members over what needs to be done. Members must have the authority and the budget allocations needed to implement the decisions made. There must be enough flexibility in the agenda of each agency to accommodate new joint tasks.

Responsiveness to the needs of targeted clients. As the experience in Zimbabwe shows, a coordinating committee can catalyze the translation of a national policy favoring a particular client group of farmers into coordinated research and extension programs.

Capacity to transfer relevant technology. Coordinating committees can be a valuable way of generating consensus and will be needed if research and extension are to cooperate in developing and disseminating a new technology. Much depends on whether both parties consider the technology has a high potential to benefit the welfare of targeted clients. The participation of on-farm researchers and extensionists on committees to approve the release of new plant varieties or modify technical recommendations to farmers can facilitate the work of both groups. This has been an effective function of COFRE in Zimbabwe. In Zambia, meetings convened to revise recommendations have been one of the few occasions which have brought subject-matter specialists from extension and on-farm researchers together.

Institutional sustainability. As a mechanism, committees are usually formal, representing some degree of institutionalization. Yet, to be sustainable, they have to be incorporated into the regular procedures and staff responsibilities of the institutions involved. Coordination committees that exist in name only are all too common in research and extension systems. Such committees also need to be flexible and dynamic. Their composition may need to change to reflect the nature of the technology currently being transferred, or the kind of information sought from farmers and extension agents. If committees become routinized, members will come to feel that membership does not contribute to their work and attendance at meetings will decline. Thus, the effectiveness of the committee as a linkage mechanism is reduced.

Participation of Extension Field Staff in Implementing
Field Surveys and Trials

Examples from the case studies. In a number of the programs reviewed in the case studies, field-level extension staff were directly involved in on-farm research, both as interviewers in surveys and as assistants in the day-to-day management of experiments. There are two ways in which this can be done: routine tasks can be delegated to extension agents in addition to their regular duties, or extension agents can be formally seconded to the research agency to perform certain tasks.

Delegation of research tasks to extension agents. Delegation is a tempting option, because it allows the geographical coverage of a research program to be increased through the use of existing extension personnel. Nevertheless, the case studies show that unless researchers work closely with them, extension agents are rarely able to manage experiments successfully. When the management of experiments is added to their normal duties, extension agents do not have the time, training, experience, mobility or motivation to keep loss rates and coefficients of variation down. It is a recipe for frustration — everyone involved ends up feeling they are wasting their time.

The problems of obtaining good data from the farmer field trials and minikit program in Nepal have already been discussed. The case study from Zimbabwe provides another good example of this problem. The Agronomy Institute, a division of the research unit of DR and SS, instituted an on-farm testing program immediately after independence in 1980. Called the Communal Areas Research Trials (CART), the program's goal was to adapt existing technology to the conditions of resource-poor farmers in the communal areas. Experiments on a range of different crops were scattered widely. They were designed by the research staff, but their routine management was left to local agents of the extension agency (AGRITEX) and to the farmers themselves. Assistants were trained at annual 4-day workshops on trial design and data collection.

It was not an effective strategy. The research scientist in charge was forced to travel constantly, but still did not have time to think through the experimental design appropriate for each site or to interact with the extension agents and farmers. Many trials were lost altogether, and few useful data were fed back into the research process. Almost no technology immediately suitable for transfer to farmers was identified. The program was reorganized in 1984 with an increased focus on applied research. A greatly reduced number of trials were clustered in a few representative areas under the direct management of technicians from the research institute who were outposted to the sites. The results became much more valuable.

Secondment of technicians from extension to on-farm research. The ARPT program in Zambia is the only instance in the case studies in which the technicians responsible for on-farm trials are formally seconded from extension. Once trained, these Trials Assistants become effective members of the field research teams. They speak the local languages, and understand local agronomic practices and food preferences. However, the mechanism has not functioned effectively as a link with extension agents outside the areas where the ARPT field teams have conducted on-farm trials.

Assessment. Research organizations need field technicians when they set up on-farm research programs far from their normal bases of operation. Extension agents can meet this need at relatively low cost, but careful management is required if they are to produce satisfactory research results and also serve as a link with the extension system as a whole.

Responsiveness to the needs of targeted clients. If extension agents are local people who speak the farmers' language and are familiar with local farming practices and constraints, their participation in the research process can increase its responsiveness. However, the experiences in Nepal and Zambia demonstrate that merely including extensionists in on-farm research does not guarantee that their knowledge and experience will actually be used in research priority-setting and planning — if this is to happen, specific feedback mechanisms to higher levels must be developed and managed.

Capacity to transfer relevant technology. Participating in on-farm research can help extensionists understand a new technology and explain it to farmers, but this is effective only if the data are analyzed and interpreted in terms of local conditions. Extensionists almost inevitably have lower status and educational levels than researchers. If this mechanism is to be effective, they must be respected as valuable team members, not used simply as cheap labor to increase the number of trials that can be run. Their direct experience with research can also help them explain results to other extensionists who do not take part directly. This influence will not extend beyond the immediate areas where research is done unless extensionists are rotated through the on-farm research program or participate in formal training courses.

Institutional sustainability. The incorporation of field staff from extension into on-farm research can be sustained on a regular basis only if their other responsibilities are reduced and if permanent funding arrangements are made. Mechanisms ensuring that information flows in both directions must be developed if the link is to improve the effectiveness of both institutions.

Participation of Senior Extension Specialists as Scientists in On-Farm Research or of Researchers as Outreach Officers in Extension Programs

Examples from the case studies. Senior extension personnel can serve as valuable members of on-farm research teams. They can facilitate flows of information in both directions: summarizing reports on farmers' conditions from local extension agents for use by researchers, and synthesizing the results of research into communications materials for extensionists to use in the field. Outreach officers from research can play analogous roles in extension programs. On the other hand, it is not easy to work in a job where responsibilities and lines of responsibility are split between two institutions.

Partial participation in Nepal and Zimbabwe. In Nepal, the British-funded Lumle and Pakhribas Agricultural Centres have their own extension programs in selected target areas. Their extension professionals participate in both the planning and analysis of on-farm research, although they are not fully integrated with field research activities. Outreach officers from the commodity programs have worked within extension programs, although this role has not become permanent.

In Zimbabwe, the cotton specialist of AGRITEX, the extension service, has his office on the experiment station of the Cotton Research Institute, which is a division of DR and SS. He participates in both research and training for the communal areas, and develops messages for AGRITEX's radio programs.

Research-Extension Liaison Officers in Zambia. These officers are fully fledged members of some of the provincial ARPTs. They are involved in a wide range of activities, including the planning and implementation of on-farm demonstrations, the organization of field days and in-service training programs, the production of regular newsletters for distribution to researchers and extension workers, and the preparation of extension materials. The divided responsibility and ambiguous job descriptions for these positions makes them difficult to fill.

Assessment. Most links between research and extension require communication between different institutions and between people of different status and educational level. The few cases where professionals from extension have been brought in to participate as equals in on-farm research programs show this to be a promising strategy.

Responsiveness to the needs of targeted clients. Senior professionals from the extension department have both the mandate and the stature to keep on-farm research programs focused on farmers' priority needs. Outreach officers from research are well placed to alert the research group to technology adoption problems encountered by extension agents.

Capacity to transfer relevant technology. Full-time specialists with a clear understanding of the structure and needs of the extension system expedite the flow of useful information and technology from on-farm research. Outreach officers from research are in a good position to synthesize experimental results into a useful form.

Institutional sustainability. In spite of these advantages, it is difficult to work for one institution and operate in another. Research and extension are parallel branches of the same organization in Zambia, the only example in which this mechanism is well developed, and even there the position of the Research-Extension Liaison Officers has been ambiguous. The long-term sustainability of cooperative participatory arrangements between research and extension probably depends on whether or not the two groups as a whole are developing shared goals and operational procedures. If they are drifting further apart, with the result that rivalry is developing between them, participatory arrangements are unlikely to survive.

Participation of On-Farm Research Staff in Rural Development Projects

Examples from the case studies. Integrated rural development projects have often sought out on-farm research programs to cooperate in the development of locally adapted technology. The advantage of these arrangements is that researchers and extensionists can collaborate closely under a single funding and management structure. However, there are some dangers. Development programs are vulnerable to frequent shifts in the goals and focus of their donors. They often ask researchers to work on whatever problems are most pressing at the moment. This can conflict with broader, long-term research goals, and make it difficult to accumulate and interpret data according to consistent criteria.

Coordination with regional development agencies in Senegal. For over 20 years, on-farm research in Senegal has included the issue of technology transfer on its agenda. Integrated research and extension programs known as *unités expérimentales* (experimental units) were designed by French researchers in the 1960s to raise groundnut yields through the diffusion of tested technology (Bingen and Faye, 1985; Fresco and Poats, 1986). This was the background for the on-farm research program set up by the Institut Sénégalais de Recherche Agricole (ISRA) in the 1980s with funding from USAID and the World Bank.

Extension services were organized within regional development agencies for Senegal's major river basins. They developed two different kinds of link with on-farm research at ISRA. The Senegal River basin authority signed contracts with ISRA for particular lines of research designed to contribute to well-defined development objectives. On-farm experiments were organized jointly by research scientists and extension agents. The trials were used as an opportunity to train the authority's field staff in farm-level conditions. In the Casamance River basin, collaboration between ISRA and the Société pour la Mise en Valeur de la Casamance (SOMIVAC) was mandated by two separate donors. USAID made the disbursement of the second phase of funding contingent upon the establishment of a formal protocol between research and extension. A liaison committee was established to implement the agreement.

Joint activities consisted primarily of regular meetings between senior researchers and senior management in the agency. These had several positive results. SOMIVAC agreed to redefine its operational zones, which had been based solely on soils and hydrographic data, using an alternative system developed by the on-farm research team which included socioeconomic criteria. Several lines of research on the local experiment station were initiated in response to needs identified by the development workers.

A major weakness was that the meetings were attended primarily by senior personnel from both agencies, most of whom were expatriate scientists. Neither field-level extension workers nor farmers were directly involved. Because the link was not institutionalized, the process of active coordination did not survive the departure of a few key individuals.

Quite separately, an appraisal of the project by the World Bank recommended the appointment of a Research-Extension Liaison Officer. The proposal was never fully discussed with either ISRA or SOMIVAC, and neither agency would appoint a person to fill the position.

Providing manpower to rural development in Ecuador. Five of the 10 regional PIP teams in Ecuador have participated directly in projects of the Programa de Desarrollo Rural Integrado (PDRI), the country's integrated rural development program. Researchers assigned by the Instituto Nacional de Investigaciones Agropecuarios (INIAP) work closely with the projects' extension staff. Farmers volunteer as collaborators at meetings convened for broader purposes by the project. The major advantage of the close association of research with other aspects of the project is that locally tested technology is provided to the beneficiaries in an integrated package of inputs, credit and advice. A disadvantage has been that, under pressure to show short-term results, on-farm research scientists have been drawn into service functions such as the multiplication of seed and the distribution of inputs. Restrictions on the projects' budgets have further reduced the range of subjects researched.

Joint management in Indonesia. The Upland Agriculture and Conservation Project in Indonesia is a regional development project managed cooperatively by the several agencies involved, including both research and extension. The research agenda of the project is designed and monitored by a technical advisory team of senior research scientists, who have identified component technologies for adaptation and testing on-farm. Extension staff are consulted in the planning and implementation of on-farm experiments as frequently as once a week. Once promising technology is identified, special training courses for field extension workers are held in the target areas. The field extension workers are then responsible for implementing pre-production verification trials and for instructing farmers on how to apply the new technology.

Assessment. All the on-farm research programs that have collaborated closely with large-scale rural development projects have experienced a tension between the advantages of more efficient links with technology transfer and support systems on the one hand, and the disadvantages of losing autonomy and being subject to the pressures of the short-term production goals of the development projects on the other. Conflicts can easily arise because of the differing goals, methods and operational time frames of the research programs and the development projects.

Responsiveness to the needs of targeted clients. Rural development projects are planned on the basis of an assessment of local conditions and needs. When they are targeted at increasing productivity on small farms, their managers often find that little appropriate technology is available. Adaptive on-farm research teams are often called in after the targets and goals have been set. This provides the on-farm research program with clear objectives, but also reduces its flexibility to develop and adjust its own agenda on the basis of its experience with farmers.

The integration of on-farm research with development projects has another cost. Almost invariably, the link between on-farm adaptive research and the applied research carried out on experiment stations weakens. On-farm research comes to be viewed as an extension rather than a research activity, and opportunities for communication and interaction become more limited. As a result, feedback on farmers' needs is inhibited, with potentially negative consequences for the relevance of applied research (Merrill-Sands and McAllister, 1988).

Capacity to transfer relevant technology. Although providing feedback to research may be more difficult in these situations, it becomes much easier for on-farm research to contribute to technology transfer. Projects provide established channels through which technology can be transferred to farmers, along with the necessary credit and inputs. Links are clearly most successful when there is technology available "on the shelf", ready for local adaptation.

Institutional sustainability. Development projects are normally funded by donors for relatively limited periods. Funds for personnel, vehicles, travel allowances and other operating costs facilitate close working relationships between researchers and extensionists. These are vulnerable to major changes in a project, or to its termination, unless special efforts are made to incorporate the linkage mechanisms into the regular procedures of the institutions involved, and unless sufficient funds are provided through regular channels.

Integrated On-Farm Research and Extension Programs

Examples from the case studies. The case studies document two types of program designed to bring on-farm research and extension together in an integrated system: production programs, and T and V extension.

In the first section of this paper, the approach of the production programs developed by IRRI through its Asian Cropping Systems Network was described with respect to Nepal. Successful progress through the research, extension and implementation stages is limited to regions with two basic characteristics: the yield potential of the major grain crops in the improved system must be high, and the distribution of the necessary inputs must be feasible.

The T and V system of extension is a highly programmed system developed in the late 1970s by the World Bank (Benor and Baxter, 1984). It has been financed and promoted in many developing countries. According to the model, village-level extension workers deliver technological messages to selected contact farmers according to a regular schedule. These farmers are expected to pass the information on to others in their area. The extension workers attend fortnightly training sessions, each of which is focused on messages appropriate to farmers' activities at the current stage of the growing season. T and V is a rigid, hierarchical system which emphasizes continuous monitoring and evaluation. Some countries have included on-farm research directly in their T and V system; others have depended on cooperation with on-farm work implemented by research institutions.

The spirit underlying the top-down structure of the T and V system is very different from that of most on-farm research programs, with their emphasis on flexible, adaptive research. Nevertheless, T and V systems create an institutional demand for locally adapted technological "messages" to present at the regular extension meetings. Several of the on-farm research programs in the case studies had developed mechanisms to satisfy this need for a constant stream of information.

A successful T and V program in Bangladesh. The most successful example in the case studies of a program of this type developing effective research-extension links through on-farm research is the Extension and Research Project of the Bangladesh Agricultural Research Institute (BARI). It was initiated in 1978 in the high-potential northwestern region of the country. Extension activities had previously been scattered between eight specialized organizations, each with its own mandate and methods. The World Bank provided substantial funding to reorganize them into a single T and V system, supported by new facilities, staff, vehicles, training and operating expenses for both research and extension.

The primary goal of the research project was to provide answers to the many questions posed by farmers and extension workers. Other objectives included delineating the areas where existing packages of improved seeds and practices were and were not appropriate, developing agronomic recommendations for local varieties, and identifying the potential for new crops within existing farming systems.

It took some years for BARI and the new extension organization to develop effective mechanisms for joint planning and coordination. In 1980, a 2-day meeting was called to discuss links between agencies, to plan the on-farm research program for the following year, and to set supply and equipment needs. It was over in less than 2 hours, because nobody present knew how to prepare or carry out an exercise of this kind. The approach used by IRRI's Cropping Systems Network was subsequently adopted precisely because it provided clear guidelines on how to proceed.

The hierarchy of coordinating committees created on paper under the T and V model never functioned, because the senior administrators named as their chairmen did not have the time or incentive to organize them. Because the researchers and extension workers felt the need to coordinate their activities, they organized their own technical committees at regional and district levels. These became important bodies which met 5 to 10 times a year.

As they gained experience, the researchers instituted a number of important innovations. They involved personnel from extension directly in site selection and diagnostic surveys, and in the design and testing of cropping patterns. They made an effort to identify innovative farmers, learn what they were doing, and pass the results laterally along to farmers in other areas. They developed flexible procedures for on-farm research which were later adopted by other divisions of BARI. At the same time, they satisfied their specialized mandate by organizing field days and training programs for extension workers, and by providing various kinds of information to the extension system.

Other experiences. The basic challenge of the T and V system is to provide enough new information to farmers to justify the cost. Experience in both Zambia and Nepal suggests that unless farmers receive concrete benefits, they become bored, refuse to be contact farmers, and stop attending meetings (Sutherland, 1986). The system has worked best in densely populated regions where production systems are relatively homogeneous, so that a single technical message is appropriate for a large number of farmers, and where the ratio of closely supervised local extension workers to farmers is high. It has been much less successful elsewhere, in part because it becomes impossible to identify enough widely appropriate technology to send down through the complex structure (Howell, 1988).

Assessment. Both production programs and T and V provide a framework for establishing links between on-farm research and extension. Both are organized hierarchically, with set roles fixed for all parties in advance. Production programs are initiated from the research side, and include mechanisms for extensionists and input-supplying agencies to carry the technology on to farmers. T and V systems are initiated from the extension side, and include mechanisms to obtain the necessary technological messages from researchers.

Responsiveness to the needs of targeted clients. The cropping systems programs in the case studies did not involve extensionists in the selection of sites or in surveys of farmers' practices and constraints. T and V systems operate within hierarchical, formalized organizational structures which emphasize the close supervision of local extension workers. Neither system facilitates feedback from farmers to researchers, either through extensionists or directly.

Capacity to transfer relevant technology. Both systems are oriented towards increasing production as rapidly as possible, and have developed a variety of linkage mechanisms to move technology to farmers. Both are successful primarily in high-potential areas with relatively homogenous farming systems. Resource-poor farmers in more heterogeneous farming systems tend not to benefit. Both are biased towards the introduction of packages of new technology with associated inputs.

Institutional sustainability. Production programs and T and V systems have been funded by external donors. Many of their linkage mechanisms depend on vehicles, maintenance, reliable travel funds for regular meetings, and other recurrent costs, as well as on a continuous supply of technical inputs and messages. Unless the usefulness of these mechanisms is clearly demonstrated, they will become vulnerable as the programs are institutionalized and unless national programs are firmly committed to meeting their operating costs.

General Lessons

Conditions for Building Effective Links

Ideally, an effective program of research and extension for the adaptation and transfer of technology to small-scale farmers should be based on the following conditions:

1. a shared analysis of target farmers' conditions and problems
2. technical alternatives to farmers' current practices which can be successfully adapted to suit local circumstances through on-farm research

3. well-trained and committed professionals in the institutions responsible for both research and extension
4. a clear division of responsibilities, assigning to each institution a set of tasks for which it has a relative advantage
5. effective linkage mechanisms, together with administrative and budgetary support, which allow researchers and extensionists to plan and carry out coordinated programs

None of the countries in the case studies met all these conditions. Only in a few cases had research and extension even attempted to organize joint activities directed towards common goals. In most cases, public research institutions had established on-farm programs on the assumption that this would overcome the most important barriers to getting improved technology to small-scale farmers. Often there was a feeling that this was necessary precisely because the extension institutions were not doing their job effectively.

On-Farm Research: No Substitute for Extension

The on-farm research programs documented in the case studies made important contributions towards improving the process of defining the needs of resource-poor farmers; it would seem that they are better suited for this role than extension services, which are sometimes biased in favor of more prosperous farmers. In many cases they also successfully adapted technology and transferred it to small-scale farmers within their immediate project area. Recommendations tailored to location-specific circumstances have been developed — a great improvement over the blanket technology packages extension services often promote.

However, the coverage of on-farm research is not broad enough. Widespread impact is limited by the chronically weak links between on-farm research and extension. The case study experiences argue forcefully that on-farm research cannot substitute for extension. Good institutional cooperation is crucial if new technology is to be broadly verified and transferred to a full range of clients.

Anticipating the Need for Links with Extension

Links with extension were a secondary priority in many on-farm research programs, and virtually all the case studies concluded that this had been a weak area in the implementation of on-farm research. Often, managers had failed to think about links with extension until technology was ready to transfer. Thus, one of the major conclusions of this comparative study is that on-farm research programs need to pay more attention to forging links with extension or other technology transfer agencies, if the process of transferring and diffusing technology is to become more effective.

Links between on-farm research and extension are likely to be more effective when they are built in at the early stages of an on-farm research effort, rather than when they are hastily created, as on-farm research produces technologies for widespread verification and demonstration. Establishing links at an early stage, while it may appear wasteful when there is as yet little technology to transfer, has two important advantages: it allows extension to contribute to the planning of research and hence increases the likelihood that research will be relevant to clients' needs; and, more important still, it means that the structures and procedures for technology transfer will be in place when they are needed — the research and extension staff responsible for linkages will be better trained and motivated, and will share a common sense of purpose. Indeed, the early establishment of linkage mechanisms may exert a positive demand for relevant technology on the adaptive and applied research system, increasing the pressures on the system to perform.

Targeting Resource-Poor Farmers

Equity was a major concern in all the case studies. The on-farm programs had attempted to develop technology appropriate for resource-poor farmers in marginal agro-ecological zones. The record was a mixture of success and failure, but it must be recognized that this is a challenging problem even in developed countries with well-established institutions. On-farm research programs as different in their philosophies as ICTA in Guatemala and the production programs in Nepal were most successful with relatively prosperous small-scale farmers working under relatively favorable conditions. Links with extension had not contributed much, in part because most extension institutions are biased toward so-called "progressive" farmers, who are in a position to adopt yield-enhancing technologies. On-farm research programs had partially compensated for this bias in the area of diagnosis and prioritization of farmers' needs.

Alternatives outside the public sector need to be explored carefully. Nongovernment organizations (NGOs) often have a long-term, focused commitment to development in poor rural areas and are less hampered by bureaucratic constraints (Sager and Farrington, 1988). In the case studies, there were several examples of successful cooperation between on-farm research programs and NGOs. In Guatemala, World Neighbors effectively transferred ICTA's adaptive research results to one area of the highlands. Once methods and procedures have been worked out on a pilot basis in collaboration with an NGO, they could be transferred to the public extension service.

The Status Problem

There is a hierarchy of prestige in agricultural science throughout the world. Maintaining effective two-way communication between lower-status field researchers in on-farm programs and their higher-status colleagues on experiment stations, even in the same institutions, was a real problem in all the programs studied (Merrill-Sands and McAllister, 1988). The gap in status between researchers and extensionists is even greater and more deeply entrenched; in addition, there is often a wider institutional boundary to cross. On-farm research programs have tended to view extensionists as implementors rather than as partners. There is little evidence that the needs identified by extension institutions played a significant role in setting the research agenda of the on-farm programs. Moreover, the emphasis on adaptive research responsive to local conditions has put new demands on extensionists without providing them either with a more efficient structure or with additional resources to carry them out effectively.

The use of Research-Extension Liaison Officers in on-farm teams, as in Zambia, is an interesting development in the search for ways of bridging the status gap between research and extension. Although their intermediate position between the two leads to organizational and personnel problems, their role as technology packagers and consolidators can provide extension with a professional contribution to make to the transfer of technology and its fine tuning to local conditions. The problems encountered in defining the role of such officers — awkwardly straddled between organizations with different objectives and procedures — shows their task to be a complex one.

One of the lessons emerging from the case studies is that, when setting up on-farm research programs, managers must not to do so at the expense of the existing extension service. The transfer of prestigious tasks or senior staff from extension to research can be demoralizing for extension programs, and thus reduce the chances of developing effective links in the future. Seconding staff from the extension service to the research program may help overcome this problem — as long as such officers are seen as still "belonging" to extension, and not as outsiders.

In the short term, managers must recognize that programs attempting to integrate the work of professionals, technicians and farmers across institutional boundaries and in defiance of status differences will encounter problems. In the longer term, emphasis must be placed on upgrading extension: more equal education, better training and more joint appointments are some of the measures needed.

Developing Linkage Mechanisms

Better ways of working together despite the difficulties need to be developed. The linkage mechanisms analyzed in the second section of this paper are a good starting point. The first two — informal contacts in the field and formal committees at higher levels of administration — are necessary first steps for any kind of collaboration; they provide a basis for communication about common goals and a framework for joint planning. The next two — secondment of junior and/or senior staff to specific research or extension programs — have a mixed record of effectiveness; they have been most successful where roles and job descriptions were realistically and clearly defined. The last two — which involve joint participation in common projects — clearly facilitate the transfer of technology, but have often suffered from unrealistic expectations and excessively rigid structures.

Links at Multiple Levels: A Key to Success

The most successful cases of integration of on-farm research and extension are those in which links have been forged simultaneously at several levels of the administrative hierarchy of the organizations involved: technicians in the field, scientists and administrators at regional level, and high-level national committees. It is clear from the case studies that on-farm research alone cannot solve the linkage problem.

When there are links at multiple levels, a strong apex management group can develop that not only combines the viewpoints of research and extension, but also has access to the structures and mechanisms needed to implement its vision. In Zambia, senior extension staff were involved in the initial planning of ARPTs. Provincial Agricultural Officers also provide administrative support and supervision in the field. This has helped to keep the ARPTs actively pursuing stronger links with extension.

It is too early to gauge the success of the Zambian experiment, but Research-Extension Liaison Officers working in the field may provide the crucial link between on-farm research and extension. Often seconded from extension, yet committed to the technology developed by the on-farm team, they are well placed to become product champions, enlisting the cooperation of the extension service in the verification stage and thus broadening the impact of on-farm research.

The Sustainability Issue

The sustainability of a linkage mechanism should be judged in the context of how well the mechanism contributes to an effective working relationship between research and extension institutions over the longer term. For example, an expatriate Research-Extension Liaison Officer who is working as part of a donor-funded project may stay in the job for only a few years, after which his/her position may not necessarily be replaced by a national staff position. Nevertheless, if he/she organizes workshops which lead to a regular program of joint planning and review, then the post will have been an effective mechanism. This kind of progress, however, requires leadership from senior management. Clear goals must be set, linkage mechanisms must be supported with the necessary resources, and incentives must be created to reward cooperation.

Acknowledgments

This paper is based on the nine On-farm Client-oriented Research (OFCOR) case studies listed in the bibliography (Avila et al., 1989; Budianto et al., 1989; Cuellar, 1989; Faye and Bingen, 1989; Jabbar and Zainul Abedin, 1989; Kayastha et al., 1989; Kean and Singogo, 1988; Ruano and Fumagalli, 1988; and Solíz et al., 1989). I am most grateful for the work of the authors, all of whom have been closely associated with the projects studied.

The issue of links between agricultural research and extension was discussed at length with the other members of ISNAR's core staff involved in the OFCOR study — Deborah Merrill-Sands, Stephen Biggs, Susan Poats, James Bingen and Jean McAllister—all of whom provided valuable comments on an earlier draft. I would also like to thank the following reviewers: David Kaimowitz, Stuart Kean, Sergio Ruano, Willem Stoop, M.A. Jabbar, Elon Gilbert and Douglas Horton. I gratefully acknowledge the careful work of Simon Chater and Kay Sayce, of Chayce Publication Services, in editing and preparing the manuscript for publication. As author, however, I remain fully responsible for any outstanding omissions or misinterpretations.

References

Andrew, C.O, and McDermott, J.K. "Farmer linkages to agricultural extension and research: A question of cost-effectivess." Farming Systems Support Project. Mimeo. Gainesville: University of Florida, 1985.

Avila, M., Whingwiri, E.E., and Mombeshora, B.G. "Zimbabwe: A case study of five on-farm research programs in the Department of Research and Specialist Services, Ministry of Agriculture." OFCOR Case Study. The Hague: ISNAR, (in press).

Benor, D., and Baxter, B. Training and Visit Extension. Washington, D.C.: World Bank, 1984.

Biggs, S.D. "Resource-poor farmer participation in research: A synthesis of experiences from nine national agricultural research systems." OFCOR Comparative Study No 3. The Hague: ISNAR, 1989.

Bingen, R.J., and Faye, J. "Agricultural research and extension in francophone West Africa: The Senegal experience." Paper presented at the Farming Systems Research and Extension Symposium, Kansas State University, 1985.

Budianto, J., Ismail, I.G., Siridodo, Sitorus, P., Tarigans, D.D., Mulyadi, A. and Suprat. "Indonesia: A case study on the organization and management of on-farm research in the Agency for Agricultural Research and Development, Ministry of Agriculture." OFCOR Case Study. The Hague: ISNAR, (in press).

Byerlee, D., Collinson, M.P., Biggs, S., Harrington, L., Martinez, J.C., Moscardi, E., and Winkelmann, D. "Planning technologies appropriate to farmers: Concepts and procedures." Mexico City: CIMMYT, 1980.

Byerlee, D., and Tripp, R. "Strengthening linkages in agricultural research through a farming systems perspective: The role of social scientists." *Experimental Agriculture* (24, 1988).

Cernea, M.M., Coulter, J.K., and Russell, J.F.A. (eds). *Research, Extension, Farmer: A Two-Way Continuum for Agricultural Development*. Washington, D.C.: World Bank, 1985.

Collinson, M.E. "FSR: Diagnosing the problems." *Research, Extension, Farmer: A Two-Way Continuum for Agricultural Development*. Cernea, M.M., Coulter, J.K., and Russell, J.F.A. (eds). Washington, D.C.: World Bank, 1985.

Collinson, M.E. "Farming systems research: Procedures for technology development." *Experimental Agriculture* (23, 1987).

Cuellar, M. "Panamá: Un estudio del caso de la organización y manejo del programa de investigación en finca de productores en el Instituto de Investigación Agropecuaria de Panamá." OFCOR Case Study. The Hague: ISNAR, (in press).

de Datta, S.K., Gomez, K.A., Herdt, R.W., and Barker, R. *A Handbook on the Methodology for an Integrated Experiment Survey on Rice Yield Constraints*. Los Baños: IRRI, 1978.

Denning, G.L. "Integrating agricultural extension programs with farming systems research." *Research, Extension, Farmer: A Two-Way Continuum for Agricultural Development*. Cernea, M.M., Coulter, J.K., and Russell, J.F.A. (eds). Washington, D.C.: World Bank, 1985.

Edwards, R., Gibson, P., Kean, S., Lumbasi, C., and Waterworth, J. "Improving small-scale farmers' maize production in Zambia: Experiences in collaboration between commodity research, adaptive research, and extension workers." *CIMMYT Farming Systems Newsletter* (Lilongwe, Malawi, 1988).

Ewell, P.T. "Organization and management of field activities in on-farm research: A review of experience in nine countries." OFCOR Comparative Study No 2. The Hague: ISNAR, 1988.

Faye, J., and Bingen, J. "Sénégal: Organisation et gestion de la recherche sur les systèmes de production, Institut Sénégalais de Recherches Agricoles." OFCOR Case Study. The Hague: ISNAR, (in press).

Fresco, L.A., and Poats, S.V. "Farming systems research and extension: An approach to solving food problems in Africa." *Food in sub-Saharan Africa*. Hansen, A., and McMillan, D.E. (eds). Boulder: Lynne Rienner, 1986.

Gilbert, E., Norman, D., and Winch, F. *Farming Systems Research: A Critical Appraisal.* Rural Development Paper. East Lansing: Michigan State University, 1980.

Gostyla, L., and Whyte, W. *ICTA in Guatemala: The Evolution of a New Model for Agricultural Research and Extension.* Special Series on Agricultural Research and Extension. Ithaca: Cornell University Press, 1980.

Harwood, R. *Small-farm Development: Understanding and Improving Farming Systems in the Humid Tropics.* Boulder: Westview Press, 1979.

Horton, D. "Assessing the impact of agricultural development programs." *World Development* (vol 14, no 4, 1986).

Howell, J. (ed) "Training and Visit Extension in practice." Agricultural Administration Unit Occasional Paper No 8. London: ODI, 1988.

Hudgens, R.E. "Subregional issues in the implementation of farming systems research and extension methodology: A case study in Zambia." Farming Systems Research Paper. Manhattan: Kansas State University, 1986.

Jabbar, M.A., and Zainul Abedin, M.D. "Bangladesh: The evolution and the significance of on-farm and farming systems research in the Bangladesh Agricultural Research Institute." OFCOR Case Study No 3. The Hague: ISNAR, 1989.

Johnson, S.H., and Kellogg, E.D. "Extension's role in adapting and evaluating new technology for farmers." *Agricultural Extension: A Reference Manual.* Swanson, B.E. (ed). Rome: FAO, 1984.

Johnson, S.H., and Claar, J. "FSR/E: Shifting the intersection between research and extension." *Agricultural Administration* (23, 1986).

Kayastha, B.N., Rood, P., and Mathema, S.B. "Nepal: The organization and management of on-farm research in national agricultural research systems." OFCOR Case Study No 4. The Hague: ISNAR, 1989.

Kean, S.A., and Singogo, L.P. "Zambia: Organization and management of the Adaptive Research Planning Team (ARPT), Ministry of Agriculture and Water Development." OFCOR Case Study No 1. The Hague: ISNAR, 1988.

Lionberger, H.F. "FSR/E in the world system of agricultural research and extension." Paper presented at the Farming Systems Research and Extension Symposium, Kansas State University, 1986.

Lipinski, D., and Rizal, M.P. "Pre-production verification trials in Nepal." Draft mimeo. Farming Systems Research and Development Division, Nepal.

McDermott, J.K., and Bathrick, D. "Guatemala: Development of the Institute of Agricultural Science and Technology (ICTA) and its impact on agricultural research and farm productivity." Project Impact Evaluation Report. Washington D.C.: USAID, 1982.

Merrill-Sands, D., and McAllister, J. "Strengthening the integration of on-farm, client-oriented research and experiment station research in national agricultural research systems (NARS): Management lessons from nine country case studies." OFCOR Comparative Study No 1. The Hague: ISNAR, 1988.

Mosher, A.T. *Technical Cooperation in Latin American Agriculture.* Chicago: University of Chicago Press, 1957.

Ortíz, R. "Transferencia de tecnología en Guatemala." Guatemala: ICTA, 1987.

Rhoades, R.E. "Farmers and experimentation." Discussion Paper No 21. Agricultural Administration Unit. London: ODI, 1987.

Rhoades, R.E. and Booth, R. "Farmer-back-to-farmer: A model for generating acceptable agricultural technology." CIP Social Science Department Working Paper. Lima, Peru: CIP, 1982.

Ruano, S. "Validación, transferencia, y dominios de recomendación." Guatemala: PRECODEPA.

Ruano, S., and Fumagalli, A. "Guatemala: Organización y Manejo de la Investigación en Finca en el Instituto de Ciencia y Tecnología Agrícolas (ICTA)." OFCOR Case Study No 2. The Hague: ISNAR, 1988.

Sager, D., and Farrington, J. "Participatory approaches to technology generation: From the development of methodology to wider-scale implementation." Agricultural Administration (Research and Extension) Network Paper No 2. Agricultural Administration Unit. London: ODI, 1988.

Singogo, L.P. "Organization and management of linkages between on-farm research and extension: Lessons from Zambia." Paper presented at the International Workshop on Agricultural Research Management, ISNAR, The Hague, 1987.

Solíz, R., Espinosa, P., and Cardoso, V.H. "Ecuador: Un estudio de caso de la organización y manejo del programa de investigación en finca de productores (PIP) en el Instituto de Investigaciones Agropecuarias." The Hague: ISNAR, (in press).

Stoop, W. "NARS linkages in technology generation and transfer." Working Paper No 11. The Hague: ISNAR, 1988.

Sutherland, A. "Extension workers, small-scale farmers, and agricultural research: A case study in Kabwe Rural, Central Province, Zambia." Discussion Paper. Agricultural Administration Network. London: ODI, 1986.

Swanson, B.E. (ed). *Agricultural Extension: A Reference Manual.* Rome: FAO, 1984.

Whyte, W.F. "Toward new systems of agricultural research and development." *Higher-yielding Human Systems for Agriculture.* Whyte, W.F., and Boynton, D. (eds). Ithaca: Cornell University Press, 1983.

Yadev, R. P. *Agricultural Research in Nepal: Resource Allocation, Structure and Incentives.* Research Report. Washington, D.C.: IFPRI, 1987.

Zandstra, H.G., Price, E.C., Litsinger, J.A., and Morris, R.A. *A Methodology for On-farm Cropping Systems Research.* Los Baños: IRRI, 1981.

6

Private Sector Agricultural Research and Technology Transfer Links in Developing Countries

Carl Pray and Ruben Echeverría

Within the context of the current ISNAR study on the nature and causes of linkage problems between agricultural research and technology transfer, and its aim to suggest ways in which these problems might be overcome, this paper has two broad objectives. These are to identify:

- characteristics of private sector links which could be used to improve links between research and technology transfer in the public sector
- changes that could be made to government policy to strengthen links within the private sector and between the private and public sectors

The paper is based on material gathered by the authors during interviews and literature surveys for a CIMMYT-sponsored study of private sector research in Latin America and USAID-sponsored studies of private sector research in Asia. From an analysis of this material, the authors present a number of working hypotheses on which to base further studies of linkage problems.

Links between research and technology transfer serve to transform farmers' needs into researchable problems and to communicate the results of this research back to the farmers. Many types of public and private institutions conduct research and transfer technology, and the characteristics and effectiveness of these links depend upon the institutions involved. Figure 1 shows the links between public and private sector research and technology transfer (*see overleaf*).

In the public sector, most research is carried out by departments within ministries of agriculture, by semi-autonomous institutions and by univer-

Figure 1. Research and technology transfer links among private and public institutions

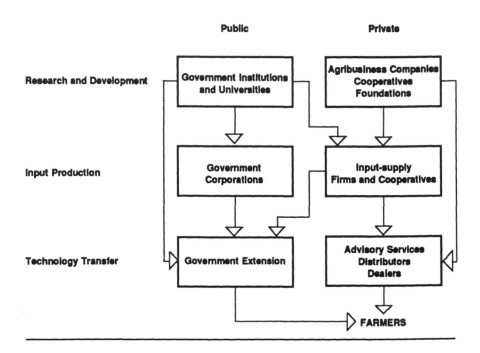

sities. Research output is embodied in an input, such as an improved plant variety, or takes the form of new knowledge, such as improved farm management techniques. Embodied technology requires an input-supply organization to produce the improved input. In many cases, input production is handled by a government corporation, which then transfers the improved input to a government extension department; in other cases, the technology is given or sold to cooperatives or private companies for further development and distribution. Management technology requires extension services to transfer the technology to the farmer. Usually, these services are provided by a government extension department; sometimes they are provided by other government organizations which are responsible for specific commodities or inputs (such as fertilizer production, sugarcane development or supervised credit). Some universities also have their own extension projects.

As indicated in other papers in this publication which focus on public sector links, the interaction between public research and technology transfer may be ill defined or inefficient. The two activities are often carried out by separate institutions which have to compete with each other for government resources and thus have little incentive to cooperate. It appears that the most effective links exist where the resources are distributed on the basis of the impact of technology in farmers' fields.

The involvement of private institutions in the generation and delivery of agricultural technology has increased rapidly in developing countries during the past 20 years. Six types of private institutions conduct research: input production and supply companies; large farms and plantations; processing companies; consulting firms and agricultural publishing companies; cooperatives and commodity groups; and research foundations. Research in input-supply companies is aimed mainly at producing technology that can be embodied in their products; the other types of institutions tend to produce technologies that are not embodied in a particular input. These products and technologies reach the farmer through private marketing and, in some cases, through government extension organizations.

For the purposes of this paper, "marketing" includes the activities of both the marketing personnel and the technical advisors of input producers, distributors and dealers, as well as the extension services of other types of institutions.

Research and Marketing in the Private Sector

Categories of Private Institutions

Each of the six types of private institutions which carry out agricultural research has different research/technology transfer links. The private agribusiness companies — input-supply companies, large farms and plantations, processing companies and consulting firms — are categorized according to what they produce: agricultural inputs, agricultural products, processed products and information. Cooperatives and commodity groups are distinct in that they are cooperatively owned by a number of farmers or companies. Research foundations are characterized by the fact that they have an independent board of directors and independent sources of funding.

Input industries. Most private sector research in developing countries is carried out by the seed, pesticide and livestock feed industries. These industries conduct mainly applied research, and within each industry this research tends to be restricted to the largest companies. Many companies concentrate only on production, and an even larger number only on distribution and marketing.

Seed and livestock feed research is conducted by multinational coorporations (MNCs) and local companies. Seed production focuses on breeding hybrid cultivars of maize, sorghum, sunflower and a few other crops. Most livestock feed research focuses on producing new materials to reduce the cost of high-quality feed. Pesticide research is conducted mainly by MNCs. They carry out the intitial research (such as synthesizing new chemicals and screening new pesticides) in Europe, the USA and Japan; screening new pesticides in field trials is conducted at their experiment stations in various agro-climatic zones, including tropical and subtropical sites; the final trials, the trials required for registration and the development of safe and effective ways of applying pesticides are conducted by local subsidiaries.

Whereas most small companies involved in research handle their own production but appoint distributors to market their products in a particular region or country, large companies usually have not only their own production division but also a marketing system that reaches farm level. These large companies, especially those which produce agricultural chemicals, also have their own technical services or extension staff who set up on-farm trials, organize farmers' field days, train sales staff and distributors and handle complaints about the effectiveness or quality of company products.

Large farms and plantations. Most research carried out by these institutions concentrates on improving management techniques. For example, in recent years Malaysian plantations have developed ways of reducing fertilizer and pesticide applications without reducing yields, and have introduced a pollinating beetle that has both cut the costs of pollinating oil palms and increased yields. Many of these institutions also develop new inputs; some of the most important rubber and oil palm varieties, for example, have been developed by private plantations.

Some large Malaysian plantations have technical advisory services that provide management assistance both within the plantation and outside it to other plantations. These technical service staff act as a communication channel between the scientists and the farm managers.

Processing companies. This category consists mainly of tobacco companies, sugar mills, brewery companies and horticultural processors. The aim of research in these institutions is to increase the productivity of the farms that supply them with the raw materials, and at improving the quality of these materials. Most research focuses on management; some is concerned with procuring new inputs, such as new tobacco and sugarcane varieties.

Processing companies have their own extension staff, who are also buyers of the crops. It is these people, rather than the marketing personnel, who have the most contact with farmers and who transmit the needs of the farmers to the researchers. Marketing personnel, particularly those in tobacco companies, provide researchers with objectives as to the quality of the product but they seldom provide information on farmers' needs. Some tobacco companies, and many companies in the dairy, processed vegetable and brewery industries, have contractual relationships with farmers whereby the company provides inputs, technical advice and credit and then buys the product at a guaranteed price.

Consulting firms and agricultural publishing companies. These institutions specialize in technology transfer, and some of them conduct applied research. Consulting firms based in the USA and Europe, such as Chemonics, Harza International, Arthur D. Little and Winrock International, play an important role in transferring technology internationally.

In some developing countries, consulting firms service the needs of large farmers. In Uruguay and Argentina they conduct applied research on cultural practices, such as fertilizer application and pastoral management, for ranches specializing in livestock and crops, and transfer information from public research stations to these clients; the information and/or the inputs are usually sold to the clients as part of a package. Consulting firms in Asia tend to concentrate on plantation crops; most of them are, in fact, the technical services departments of these plantations. A growing number are independent of a plantation base and are staffed by ex-plantation managers and technicians, but these firms conduct little or no research.

Companies that publish agricultural magazines play an important role in the transfer of technical information from the public and private sectors to farmers and from one farmer to another. Examples of such magazines are the Latin American publication *Agricultura de las Américas* and the Asian publication *Agricultural Mechanization*. These publications depend heavily on advertising from the input-supply industry.

Cooperatives and commodity groups. Many of these institutions conduct research. Their members are usually large commercial farmers, plantation owners or processors. Examples of cooperatives and commodity groups which have their own research programs are SUL and CALNU, the Uruguayan wool and sugarcane associations; FEDEARROZ and CENICAFE, the Colombian rice and coffee federations; CEPLAC, the Brazilian cacao research organization; the vegetable oil mills association in India; and the Davao banana plantations in the Philippines.

An example of a smaller organization is the Consorcios Regionales de Experimentación Agropecuaria (CREA), in the southern part of Latin America, which consists of small groups modelled on French farmers' associations. Each group is located in a specific region, hires experts to provide technical advice on farm management and commercialization issues and conducts applied research as part of regional and national programs.

Foundations. There are foundations in Latin America that conduct or fund research and provide technical assistance to farmers in the Dominican Republic, Ecuador, Honduras, Peru and Venezuela. In Asia, a number of small foundations conduct research; an example is the Tata Energy Research Institute in New Dehli, India, which conducts research in agricultural biotechnology.

Characteristics of Private Sector Research and Marketing

The goal of private agribusiness companies is to maximise profits. Thus, to produce new technologies, they must allocate funds as efficiently as possible to the various activities involved — research and development (R & D), marketing, production and others.

To allocate the right amount to R & D, these companies need information from scientists and engineers on what new products can be produced, the cost of R & D, and the probability that the products will be developed within a certain time. They also need estimates from their marketing personnel on future prices and sales of the products.

To allocate the right amount to marketing, agribusiness companies need financial information on the impact of a dollar spent on marketing on prices and sales of new products. They also need technical information from scientists about the products they intend marketing; for example, information that pests are likely to develop resistance to a new pesticide in five years or that disease resistance in a new maize hybrid is likely to break down in a few years will affect the amount of money companies consider should be spent on advertising these products.

The economic value of links between research and marketing is that they provide information which reduces the amount of research resources wasted on developing products for which there is no market and the amount of marketing resources wasted on advertising and transferring new technology. The costs of linking research and marketing relate to the time staff spend in informal linkage activities and the salaries and facilities for staff engaged in formal linkage activities. In the case of agribusiness companies which rely on other companies to provide market information and to sell their products, the costs relate to the expenses incurred in buying this information and educating the dealers or technology transfer agents in these other companies.

The research, development and production process. The two characteristics of the R & D process which distinguish it from all other activities in an agribusiness company are that it may take many years to complete and its outcome is very uncertain. For this reason, companies periodically reevaluate research projects to establish whether they are still technically and commercially viable. The process of researching, developing and producing new technology can be divided into six stages (Booz et al., 1987):

Exploration: Searching for product ideas which meet the company's objectives

Screening: Determining on the basis of a quick analysis which ideas are pertinent and merit more detailed study

Business analysis: Expanding the idea into a concrete business recommendation, including product features and a production program

Development: Transforming the idea into a demonstrable and producible product

Testing: Conducting the necessary commercial experiments to verify earlier business judgements

Commercialization: Launching the product in full-scale production and sale, thus committing the company's reputation and resources

As an idea for a new product moves from the exploration stage to the commercialization stage, costs are low at first and then rise rapidly. As shown in Figure 2 (*see overleaf*), a very small share of the total cost of a new product is spent at the screening and business analysis stages, while the largest share is spent at the commercialization stage. This provides managers with the incentive to identify and eliminate doubtful product ideas as early as possible; thus, most ideas are eliminated at the screening or business analysis stage. A study of US agribusiness companies showed that, on average: a firm reduces its product ideas from 60 to about seven at the screening and business analysis stages; about a third of the ideas which then go through the development, testing and commercialization stages are commercial failures; and the time and money spent on the many ideas that are eliminated in the early stages and on those which fail after introduction account for about 70% of total cost (*see* Figure 3 *overleaf*).

At every stage of the process, marketing plays a key role in setting priorities. In most companies, marketing personnel (along with scientists and others) contribute ideas during the exploration stage; they are involved in eliminating ideas at the business analysis stage; they play the major role in the testing stage; and they manage the commercialization stage.

Throughout this process, R & D personnel communicate technical knowledge about a new product to marketing and production personnel. If the business analysis and testing are done internally, educating marketing personnel about the product is well under way before the product reaches the commercialization stage, at which point more marketing personnel have to be educated about the product through literature and formal training courses.

Figure 2. Cumulative expenditure and time required to develop new products

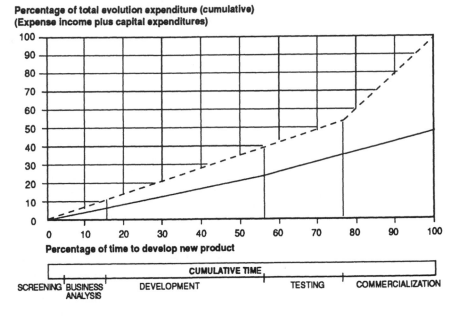

Percentage of total evolution expenditure (cumulative)
(Expense income plus capital expenditures)

Percentage of time to develop new product

Source: Booz, Allen and Hamilton, 1987

Thus, the two linkage functions — information to marketing and feedback to scientists — are going on throughout the research, development and production process.

The activities which are related to these functions take place within the company, or some of them may be contracted out to other companies. For example, some of the small biotechnology companies in the USA hire consultants to assess the market for their proposed product and they then contract large pharmaceutical or food companies to commercialize the product.

Three types of links are important:

Informal links: The norm in small organizations in which there is little need for specialized divisions; research, marketing and technical assistance personnel meet almost daily

Formal links: Common in large companies; examples include the strategic planning departments of the training departments which characterize such companies

Figure 3. Mortality of new product ideas

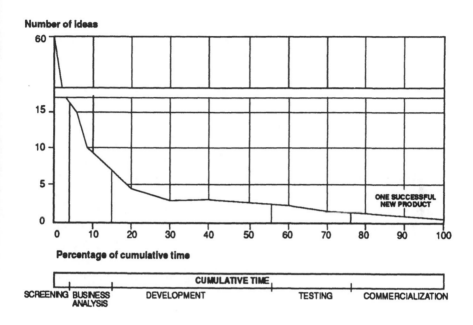

Number of Ideas

Percentage of cumulative time

CUMULATIVE TIME

SCREENING   BUSINESS   DEVELOPMENT   TESTING   COMMERCIALIZATION
            ANALYSIS

Source: Booz, Allen and Hamilton, 1987

Market links: Occur when R & D and marketing are conducted by different companies, and the service is provided for a fee

As one moves from informal links to formal links, and then to market links, the expense and difficulty of linking research and technology transfer increase. Formal links are more expensive and communication more difficult than informal links because the specialization which gave rise to the formal links is in itself expensive and it leads to status, institutional and perhaps even geographic barriers to communication. Market links are more expensive and difficult than links established within a company because, in addition to the problems of specialization, there is the problem of protecting proprietary information and this further restricts the flow of information.

The Booz, Allen and Hamilton model is applicable to small companies. They may contract out for market information and commercialize their products in joint ventures, but they can cut costs and increase the number of successful innovations by involving in-house marketing personnel or by contracting market support in the early stages of R & D. If they do not do this, they will spend more money on R & D, have fewer successful products and be less competitive.

Private research in developing countries. Private agricultural R & D in developing countries has grown rapidly in recent years. However, on the basis of the limited amount of detailed data available, it is estimated that in most of these countries expenditure on private R & D rarely exceeds 10% of the total national expenditure on research. In comparison, it is estimated that in the USA 66% (representing US$ 1.7-2.6 billion) of all research on food and agriculture is conducted by private companies (Crosby, 1986).

The most detailed data available on private R & D in developing countries derive from a survey in the mid-1980s of R & D companies in Asia (Pray, 1985; Ruttan and Pray, 1987). Most large Asian agribusiness companies with formal R & D programs were contacted (the survey did not include research conducted by commodity groups, cooperatives, consulting firms or foundations; during the colonial period most of the research was carried out by commodity groups, but since independence most of these groups have been taken over by governments).

As shown in Table 1, the data gathered during the survey were divided into five categories of private research: seeds; pesticides; machinery; livestock; and processing and plantations. The first four are input industries. Processing and plantations were placed in one category because most of the processing industries that were conducting their own research also had their own plantations. R & D by input industries makes up 60% of the private R & D in Asia. Only in two countries, the Philippines and Malaysia, does the total expenditure on private research exceed 10% of public research expenditure.

Private research in Latin America at least equals, or may even surpass, the amount done in Asia. The input industries are particularly important (de Obschatko and Piñeiro, 1985; Echeverría, 1989). Most of the major MNCs and some large local companies invest in R&D in Brazil, Argentina, Mexico and Chile. In Central America and northern South America, some plantations invest in R & D. Farmer cooperatives and commodity groups take a far more active role in R & D in Latin America than is the case with their counterparts in Asia or Africa.

Much less private research is conducted in Africa than in Asia or Latin America (Eicher, 1984; Hobbs and Taylor, 1987). In a few countries there are private research programs on oil palm, rubber and tea plantations. Private research on maize and sorghum plant breeding and, to a limited extent, on pesticides is conducted in Egypt, Côte d'Ivoire, Kenya, Nigeria, the Sudan and Zimbabwe.

In general, private research in developing countries tends to be applied in nature. International seed companies develop new hybrids by crossing

Table 1. Private sector research expenditure In seven Asian countries, 1985 (US $1 000's)

	India	Philippines	Thailand	Indonesia	Malaysia	Pakistan	Banglad...
Seeds	833 (8)ᵃ	1 583 (4)	665 (5)	0	0	182 (3)	Less th... 1 000
Pesticides	3 500 (20)	1 170 (8)	887 (5)	800 (1)	500 (3)	387 (5)	40 (2)
Machinery	6 775 (3)	none	none	none	?	none	none
Livestock	2 275 (3)	500 (6)	1 725 (2)	600 (3)	?	none	none
Processing and Plantations	3 324 (25)	1 137 (7)	1 034 (3)	600 (3)	10 000 (9)	234 (2)	50 (1)
Total Private Research	16 707	4 390	4 311	2 000	10 500	804	90
Govt Agricultural R & Dᵇ	248 000	7 000	73 595	6 700	44 400	56 170	8 000
Private as % of Govt Research	7	63	5	3	24	1	1
Agricultural Value Added ($ billions)	59.7	8.7	5.6	21.1	6.6	6.6	6.7
Private as % of Agricultural GDP	.03	.06	.05	.01	.17	.01	.01

ᵃ Number of firms in parentheses. ᵇ These numbers are not consistent in their inclusion or exclusion of capital expenditures; the Philippines capital expenditures, but Pakistan does, and in some other cases it is unclear whether or not capital expenditures have been included.

their elite lines with local germplasm. Local companies use the results of public research to develop hybrids. Multinational pesticide companies screen new products for efficiency at their experiment stations located in various agro-climatic zones, including tropical and subtropical zones. Local subsidiaries carry out the final field trials and the trials required for registration. Most agricultural machinery research takes the form of experimentation by implement producers, who incorporate modifications suggested by their own staff and by farmers.

Marketing in developing countries. The amount of money spent on marketing varies between industries and between types of companies. Much of this expenditure is not, in fact, related to transferring technology or educating farmers about new technologies but is simply product promotion.

No data have been found on the amount spent overall by agribusiness companies on marketing in developing countries. It is likely that this amount is exceeded by the amount spent on public extension services, but as countries become more developed the expenditure on public extension will probably decrease while expenditure on private marketing will increase (de Andrade Alves, 1984). In comparison, private marketing is predominant in developed countries, whereas public extension is relatively limited.

On the basis of data gathered from Asian agricultural, chemical and seed companies, it is estimated that input industries that conduct R & D in Asia spend at least two or three times as much on marketing as they do on research. In addition, there are a large number of companies that do little or no research but have substantial marketing departments. The Indian seed industry is an example: about 3% of sales of the main seed companies is spent on research, whereas about 15% of the market price goes to the distributor, of which 10% is supposed to be passed on to the dealers (Pray et al., 1989). This 15% represents the main marketing cost of Indian seed companies, but they also spend money on advertising and on training their distributors and dealers.

Plantations, processing companies and cooperatives "market" the results of their research primarily through their own technology transfer or advisory services. They do not have the costs of advertising their new products or management techniques, which reduces their total marketing budget relative to input industries. However, in most cases, the amount they spend on technology transfer probably exceeds that spent on research. Processing firms such as cigarette companies have many more technology transfer agents than researchers (although these agents are also buyers of tobacco leaf from farmers). In Pray's Asian survey, the only firms which invested more in research than marketing were the Malaysian plantation companies; they spent about twice as much on research as they did on advisory services.

Private marketing includes most of the techniques which are used by public extension: farmers' meetings; demonstration plots; short courses for farmers, distributors and dealers; radio programs; and information bulletins. There are also a number of actvities not commonly used by public extension, such as advertising on television and radio and in newspapers and magazines.

The amount of technical information that companies provide with a particular input depends on at least two factors:

- the complexity of the input in terms of management and farmers' safety (for example, more information is required on the safe and effective use of pesticides than on new crop varieties)
- the structure of the market (for example, pesticide companies selling their own product through their own distribution system invest more in providing technical services to ensure its proper use than if they sell it to the government for distribution through public extension services; companies may also invest in advertising to maintain their market share in a product if there are no property rights, such as patents, or if a patent expires)

Links between Research, Marketing and Farmers

This section describes the informal, formal and market links in R & D planning and the dissemination of research information in the private sector, and then briefly outlines regional differences in private research-marketing links.

Research and development planning

Informal links. The seed companies in India, Guatemala and Mexico provide examples of the way informal links help determine how much and what type of research is done. In India, until the early 1980s local companies were engaged mainly in marketing hybrid seed produced by public research. These companies are now investing in hybrid research because of the profitability of a pearl millet hybrid and a fodder sorghum hybrid. It was only when the marketing personnel were convinced of the profitability of research than the companies started plant breeding. Marketing personel and researchers are in regular contact with each other, and thus no formal linkage mechanisms are needed.

In contrast, the Mexican and Guatemalan maize seed industries were started by companies with a heavy emphasis on research. There seem to have been three well-defined phases in the development of research-marketing links. In the first phase, marketing had little influence on research priorities; most resources went into breeding and/or adapting hybrids from elsewhere, while marketing simply tried to sell what the researchers produced. In the second phase, marketing was concerned with the collection of basic data on seed prices and quantity and market share; there was a two-way flow of information between researchers and marketing personnel, with both groups playing an equally important role in marketing and research planning decisions; and companies were still small enough to allow the two groups to have daily contact with each other. In the third phase, marketing had developed to the point where it set the research priorities, asking researchers to develop specific products to meet farmers' needs in a particular region.

In many small agricultural machinery shops, the "researcher" is actually the owner, who is also in charge of production and marketing (Mikkelsen, 1984). He spends part of his time tinkering with the machines he produces in order to improve them, often on the basis of suggestions made by farmers, and to find cheaper ways to produce them.

Formal links. In large companies, formal links take the form of strategic planning programs and decision-making procedures that bring together research and marketing. Major decisions on important research issues (such as whether, in the case of an agricultural chemical company, research should concentrate on herbicides and fungicides and drop insecticide research) are made by top management, in conjunction with the strategic planning department. Strategic planning groups usually include both marketing personnel and technical scientists. In US-based MNCs, top management consists mainly of non-scientists.

Formal planning procedures are also used to help top management of agricultural chemical companies decide on such as issues as what type of biotechnology research to invest in. For example, once the US-based MNC DuPont decided to invest in agricultural biotechnology, a DuPont team of scientists and marketing personnel surveyed a large number of agricultural scientists and farming experts to assess which products they expected to be affected by biotechnology and when. In conjunction with advice given by scientific and marketing experts, this information was then used to determine biotechnology priorities.

Multinational agricultural chemical companies decide which crops and pests should be included for screening the biological efficacy of new chemicals by assessing the potential economic importance of controlling the pest. To do this, marketing personnel assess major markets by using public data on market size, purchasing market information from outside the company, and conducting surveys among relevant experts. Because the R & D costs of producing new technologies are so much greater than the costs of screening and business analysis, marketing personnel and information play a key role in determining which products should be developed.

R & D projects undergo regular reviews to ensure that progress is being made towards producing a technology that will be viable for the company. For example, if in the R & D process it is discovered that a chemical is effective only against insects, whereas the company is specializing in herbicides and fungicides, development might be stopped and the chemical, if the company had patented it, might be licensed to another company. If a herbicide is developed that is found to be effective against a weed that is a problem only in a small area, further development work would probably be stopped unless the government provided subsidies to meet the costs of the final development and government approval processes.

MNCs may have difficulty in overcoming distances and cultural differences between headquarters and subsidiaries. The top management of a US company, for example, decided incorrectly not to commercialize a particular herbicide in Thailand because it considered that the market was not large enough to justify costs; the head of the company's Thai subsidiary, however, was able to reverse this decision by using local funds and frequent communication with headquarters to prove that the demand was higher and the commercialization costs lower than headquarters envisaged.

In many companies, before funds can be allocated for research on a new product which accords with the general goals of the company, scientists have first to convince production and marketing personnel that the outcome has a good chance of being not only technically feasible but also profitable. At Hindustan Lever in India, for example, various profit centers in the company (such as animal feed, or plantations) decide what type of research they need, and then provide money to the central research facility in Bombay to carry out that research. Ongoing research programs are reviewed at least annually to establish whether they still serve the purposes of the profit center and, if they do not, they are stopped.

Private companies are willing to invest large amounts of money in formal linkage mechanisms to ensure that they have the information necessary to set research priorities. A good example which highlights the differences between private and public investment in this sphere concerns the Virginia tobacco breeding programs in Bangladesh. One program is run by the government, the other by the Bangladesh Tobacco Company (BTC). Both entities are aware that smoking quality is the key factor in the profitability of a new variety. The BTC built a laboratory for testing the quality of its varieties, incorporated this information into its plant breeding program and developed a high-quality variety which is popular with farmers and cigarette producers. The government, on the other hand, has not built a quality-testing facility and continues to produce new varieties which have low leaf quality, fetch low market prices and are rarely grown by farmers.

Some of the largest companies have found ways to shorten the "distance" between researchers, marketing personnel and farmers. At Pioneer Hi-Bred in the USA, the seed producer/distributor is a farmer himself. He feeds back information directly to marketing or production personnel who, in turn, pass this information on to researchers at regular meetings held between marketing and research personnel. Plant breeding stations are scattered throughout the important maize-producing regions of the USA, and the plant breeders at these stations, who are influenced by their close contact with the farmers, work with the scientists at headquarters to set research goals. In Latin America, this is taken one stage further in that in many cases Pioneer's plant breeders are farmers themselves.

Product development personnel from agricultural chemical companies and private sector plant breeders regularly work with the marketing personnel in their companies to conduct trials of new products on farmers' fields. The companies then invite local farmers to field days at which the farmers and plant breeders discuss desirable improvements in the varieties. The farmers' comments are taken into account in the conduct of subsequent trials. This pattern of using both internal and external sources of information is exemplified by the way Northrup King sets its research priorities (*see* Figure 4). Seedstock (inbred lines of hybrids and pure lines of varieties) is produced on the company's farms; the farm managers inform researchers whether an experimental variety is commercially producible; and commercial seed is produced externally by contract farmers.

Market links. These links are more common among smaller companies than in larger concerns. New companies made up primarily of scientists have to link R & D with marketing personnel by hiring marketing services rather than building internal links in the company.

In the USA, the expansion of private R & D in biotechnology was led by scientists from the private or public sector who invested capital in setting up small research companies. The companies received some guidance from the investors, and to obtain the market information they required they hired consulting firms. Many of these companies are now approaching the commercialization stage. Those that started with a strong marketing contingent in top management seem to be doing better than the science-driven companies which did not have close links with marketing at the early stages of product development.

Many companies, including both MNCs and small local firms, sell their products through other companies. By doing this, they may lose access to information on which to base research and development planning. For example, Monsanto, which concentrates on product development and manufacturing, sells its products in Southeast Asia through other companies. In the case of the Philippines, Monsanto products are sold through Bayer; Monsanto has product development staff who work with Bayer and

Figure 4. Research product feedback

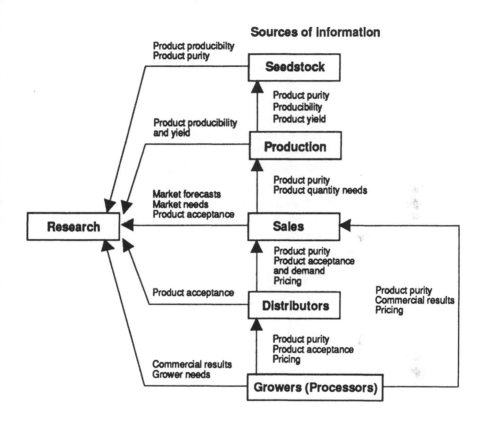

Source: Northrup King Research Department (pers. comm.), 1985

thus are in a position to provide feedback which would help set future research priorities, but because Monsanto and Bayer are rival companies in many other parts of the world, there are limits as to how much information Bayer will pass on to Monstanto and vice versa.

Companies in developing countries often buy information for use in their R & D planning activities. They buy technical expertise by hiring scientists from universities, government research programs and consulting firms as consultants and, in countries where relatively sophisticated market research firms exist, they may buy marketing information.

Dissemination of research information

Informal links. In small firms in which the only links are informal ones, the dissemination of information about a new technology is inseparable from the research process itself. In small agricultural machinery shops in the . Philippines and Thailand, for example, product and marketing personnel spend some time on research, and thus no special communication links are needed. Similarly, in small seed or tobacco companies, the dealers, technical advisors and purchasing agents conduct field trials; thus the only technical information needed is on how to conduct the trials and record the results.

If a firm is large enough to have some specialized personnel, information may still be disseminated through informal links. When marketing or technical services personnel in such companies face a technical problem, they can go directly to the scientists to discuss it.

Formal links. Large input-supply companies have several formal mechanisms for transferring research information to their production and marketing personnel. These include regular meetings at which scientists report their results to technicians and marketing personnel, and regular training sessions, often held just before each crop season, at which scientists explain new technologies or farm management practices to marketing personnel.

Another type of formal research-marketing link used by many companies is on-farm research and/or demonstrations. Seed companies' regional marketing personnel conduct trials of the hybrids in the final testing stages on rented land or farmers' fields, and in this process they acquire information about the new technology. Scientists also work with distributors and dealers to set up demonstration plots of proven new technology. Agricultural chemical companies, processing companies and cooperatives use experiments and demonstration plots to bring together researchers, marketing personnel and technical advisers.

Companies also disseminate research information through internal newsletters and, in some cases, electronic-mail networks. For example, Cargill has an electronic-mail network through which managers throughout the world report on improvements they have made in their feed mills or on new, less costly materials for producing feed. Cargill's central research department uses this network to disseminate information on its results and on public research results around the world.

Many large companies incorporate into their recruitment and personnel policies the recognition that personal communication may be the most effective way of facilitating the dissemination of research information. One way of encouraging this type of communication is to recruit marketing personnel whose technical background and experience is such that they will not feel intimidated by the scientists and will feel free to ask them for advice. Another strategy is to move, when possible, personnel between R & D, technical advisory and marketing roles. MNCs constantly move scientists and technicians between headquarters and subsidiaries in developing countries. Scientists from headquarters often visit the subsidiaries, carrying the latest information from the central research department and bringing back information from the developing countries; likewise, scientists and technical personnel from the subsidiaries visit headquarters to exchange technical and marketing information.

Market links. When market links are the main channel through which research information is disseminated, companies have to train not only their own marketing and technical personnel but also the personnel in the companies to whom they sell new technology. Input companies in Asia train thousands of dealers each year. In Bangladesh, for example, Ciba-Geigy conducted a 3-day training course in basic agriculture and the use of pesticides for 2000 pesticide dealers in 1985.

Regional differences. In Latin America, the local private sector is less actively involved than the MNCs in R & D and marketing. The production of chemicals (herbicides and pesticides) is done almost entirely by MNCs. However, a few local companies have played an important role in some areas (wheat seed in Argentina, agricultural machinery in Argentina and Brazil, rice and coffee in Colombia and cacao in Brazil).

Within Latin America there are regional variations in the links between research and marketing. In countries with large markets, such as Argentina, Brazil and Mexico, it is common to have foreign and local companies doing both research and marketing. There is less research in countries with smaller markets, such as Bolivia, Ecuador, Paraguay and Uruguay, and the predominant research-marketing links are those between R & D headquarters (regional, or in the USA or Europe) and local marketing subsidiaries; there are also a few foreign and local companies in these countries which specialize in marketing public research results.

Much of the private research effort in Asia is conducted by large local companies, and thus there are more research-marketing links within local companies than is the case in Latin America. In Africa, some countries, such as Côte d'Ivoire, have strong commodity organizations involved in both research and marketing, as well as a few private input-supply companies which are regulated by the government.

Relationship between the Private and Public Sectors

Links between Private and Public Research and Technology Transfer

Governments intervene in the provision of new technologies for several reasons. Firstly, they act in response to market failure. When private firms are unable to make a profit from investments in R & D and marketing, they begin to underinvest in these activities; governments respond by investing in research or implementing policies that increase private sector incentives to invest. Secondly, they intervene to keep markets competitive and protect local producers from foreign competition. Thirdly, they intervene to improve income distribution in the country. Fourthly, they regulate industries through legislation designed to safeguard health and protect the natural environment. In addition, governments may implement policies, such as import barriers, which, although not specifically aimed at R & D and technology transfer, do have an important impact on these activities.

The most important type of government involvement in the provision of new technologies is public sector research and extension. In Latin America, research and technology transfer was dominated initially by public organizations; this was followed by a period when private sector R & D and marketing grew substantially, especially in agricultural chemicals and machinery and improved seed varieties.

In Asia, private R & D and marketing started from a small base and has grown quite rapidly in recent years, but public research and extension continues to be predominant. Thus links between public research and private marketing are more common in Asia than in Latin America. In Africa, private R & D and marketing is very limited; only in a few countries has there been any growth in private sector involvement.

Public and private sector research links. The interaction between private and public sector research usually takes place at the individual level. This is mainly because scientists from both sectors have often studied at the same university, the number of scientists working on a specific project is small, and many private sector researchers worked initially in the public sector.

In addition to these informal channels of communication, there are some formal links, such as publications in scholarly journals, professional society meetings and meetings to set public research priorities. Cooperative research projects are also a form of linkage mechanism; for example, companies may provide a new technology (in most cases, seeds or chemicals) to be tested by public experiment stations, and the discussions of the results of these tests provide a good opportunity for interaction.

Market links between public and private research consist of contract research by public institutions and public sector scientists working in private institutions as consultants. In India, several large agribusiness companies contract researchers from university agricultural departments or from management institutes. In Southeast Asia, many university scientists work as consultants on private research projects. In Latin America, government scientists often do part-time research for private companies.

Public and private sector technology transfer links. Informal links between public extension agents and private company marketing personnel are less common than in the case of public and private sector scientists. Among the reasons for this are differences in background, training, the size of the region covered and the type of farmer with which each sector works.

In some countries there are formal arrangements for cooperation between the two groups. Private companies sometimes provide government extension departments with training in the use of new inputs. Government extension agents may arrange meetings between farmers and private marketing personnel or technical advisors.

There are also some examples of market links between the two groups. In several Southeast Asian countries it appears that some government extension agents "moonlight" as salesmen or demonstrators for private input-supply companies. There are also reports that in some developing countries distributors from private input-supply companies pay government extension agents to push their products.

Private sector transfer of public sector technologies. There are many cases in developed and developing countries where public research results are transferred to farmers by private companies. Local companies which do not have research programs depend on technologies developed by public research; they market public research results and compete with larger companies which have their own research programs.

In many developing countries, public research improves germplasm and develops hybrids and other varieties which seed companies then commercialize. In India, the flow of seed from the public to the private sector involves using informal, formal and market links. An example of the use of informal channels is when companies receive seed samples from friends or relatives who work in public research institutions. Companies also acquire new varieties through formal channels; for example, coordinators attached to the All India Crop Improvement Project (AICIP) may release seed to private companies upon request; most Indian public research institutions, as well as the International Crops Research Institute for the Semi-Arid Tropics (ICRISAT), provide breeder seed free of charge.

Seed companies also use market links to acquire new varieties developed by public research. A few universities in India and the Kasetsart University in Thailand sell their maize inbreds to private companies. Companies also buy foundation seed from the National Seed Corporation and the State Seed Corporation in India. Local seed companies in Guatemala buy maize foundation seed from the public research institute and pay a royalty for basic seed developed by the institute.

Public sector transfer of private sector technologies. The transfer of privately developed technology by public extension agents occurs when these agents recommend and/or sell privately developed inputs. An example of formal private research-public extension links can be found in the US dairy industry, when public extension agents recommend a new milking parlour which will use privately developed milking equipment; private companies, such as Alfa-Laval and Surge, set up demonstration farms where they explain the advantages of the new equipment to the extension agents. These companies may also provide equipment to universities for testing, and send information about their products to extension agencies.

Farmers frequently ask public extension agents for advice about inputs which have been privately developed. When public extension is part of a government credit system, private sector inputs are usually included in the credit and extension package.

Market links are found in the input distribution systems in many developing countries. Fertilizers, seeds and pesticides developed by private companies are imported and/or purchased locally by the government and then recommended and sold to farmers by the extension agents.

Effect of public research and extension on private sector links. Private R & D activities are influenced by public sector research in several ways. Private sector scientists get new ideas and inputs from government scientists when the two groups meet informally or formally, such as at conferences. The results of projects undertaken by the public sector to compare the usefulness of similar technologies can help shape private research programs.

The existence of an effective government extension system can influence research and marketing, and the links between them, in private companies. A company may change its research priorities to obtain government approval; for example, a private research company in India may aim to produce a variety which stands a chance of winning the AICIP yield trials, although producing seed from this variety is not economically viable. Companies will also formulate marketing strategies aimed at convincing government scientists and extension agents about the effectiveness of their products, and thus the information that flows between a company's research and marketing personnel will have more to do with what extension wants, rather than what farmers need.

Government procurement of agricultural inputs affects the types of links between private research and marketing, and the information flowing through these links. If government purchases account for a large share of the market, private input companies have to devote some of their time to predicting and influencing government demand rather than meeting the needs of commercial farmers. Thus, instead of using technical marketing personnel who are skilled in communicating with farmers, companies hire people who have government connections and are able to communicate effectively with government personnel.

In some countries, government ministries or extension systems have encouraged the establishment of input industry associations or commercial farmers' organizations. Among the tasks which industry associations perform are educating the public about the use of new inputs and gathering and disseminating information, some of which is useful to individual companies in helping them determine their research priorities; thus, these associations perform some of the linkage activities which otherwise would have been undertaken within a company, or not at all. For example, the Fertilizer Association of India collects and publishes a considerable amount of agricultural data, and has played an important role in popularizing the use of fertilizers through setting up thousands of demonstrations throughout the country.

Impact of Government Policies on Private Sector Links

The structure and efficiency of links in the private sector and between the private and public sectors are affected by how much private research and marketing government policies and regulations allow or encourage, their effect on company decisions as to whether to use informal, formal or market links, and the manner in which regulations are executed by government policy makers, scientists and extension agents. The policies and regulations which have most impact on private sector links are those concerned with: government approval of new technologies; import restrictions; restrictions on MNCs; property rights; price controls on new technology; and tax incentives for R & D.

In Table 2 (*see overleaf*), an assessment of the impact of these measures is given in columns 2 and 3. (For example, import restrictions on a commercial product which embodies new technology, such as a pesticide, would protect local companies and might encourage them to conduct more research — hence, the plus sign in column 2; such protection would, at least initially, have a negative effect on technology transfer in that the product would not be available to farmers — hence, the minus sign in column 3.) The role of public research in making or enforcing policies and regulations is given in column 4. (Using the example of import restrictions, public sector scientists do not have a major impact on policy but may be asked for advice on which imports are particularly important to the country's agriculture.)

Government approval of new technologies. The regulation of new technologies can be seen as a linkage activity in that it determines which technologies should pass from the R & D to the marketing stage. In several countries, privately developed seed varieties must be approved by the government before the company is allowed to sell the seed. Government

Table 2. Impact of government policies and regulations on private research and marketing

Policies and Regulations	Impact on Private Research	Impact on Private Marketing	Public Research Role
Govt approval of new technology			
Seed certification required	−	−	Run tests/approve
Registration of chemicals	−	−	Run tests/approve
Import restrictions			
Commercial products	+	−	Advise
Research inputs	−	−	Advise
Restrictions on MNCs			
Research permission	−	−	Advise
Foreign ownership	−	−	None
Repatriation of profits	−	−	None
Property rights			
Patents	+	+	Little
Plant variety protection	+	+	Oppose
Price controls on new technology	−	−	Advise
Tax incentives for research	+		Approve

Note: + positive impact; − negative impact

field tests are conducted to ensure that a new variety is resistant to important diseases and produces higher yields than other commercial varieties. On the basis of these tests, a government-appointed board decides whether the new variety should be released; these boards usually include public sector plant breeders, some of whom may have their own varieties in the trials, and thus the decision may be a biased one. The process involved in obtaining government approval of a new commercial variety may take several years.

Companies that introduce new pesticides are required to show that these products are effective and meet certain toxicology and environmental standards; these requirements are almost universal, but their enforcement varies greatly from one country to another. Public institutions are called upon to test the effectiveness of the product against pests, and government scientists may be asked for their advice on whether or not the product should be approved.

The process of approving new technologies brings the public and private sector together during the testing stages, the formulation of new regulations and the publication of results. This may lead to more informal contacts and better communication between the sectors; in some cases, however, it may reduce communication (for example, where there is conflict of interest, or an adversarial relationship develops between the two groups). If government trials to test new technologies are well organized, they can influence the direction of private research. By testing a new technology and then publicizing the results, governments can encourage companies to compete on the basis of the product itself, rather than on the basis of advertising.

Government regulations may also alter the links between private sector R & D and marketing. For example, to meet regulations concerning the toxicology and environmental impact of pesticides, more scientists may be required; this prevents small firms from entry into the industry. Such regulations may also influence company decisions on the type of marketing personnel to recruit, and may require than marketing personnel focus much of their attention on working a product through the regulatory system.

Import restrictions. Governments in many developing countries impose import restrictions on inputs which embody new technology. These restrictions range from a total ban on importation to the introduction of tariffs. Some countries have limits on the levels of royalties that can be paid to import new technology. The Philippines Board of Investment, for example, tries to keep royalties below 2% of sales, while India limits royalties to less than 1% of profits.

If import restrictions encourage local companies to undertake R & D, rather than relying on foreign technology, this may strengthen links between R & D and marketing because the two activities would be taking place in the same countries and thus there would be fewer cultural barriers to communication. However, import restrictions may also have the effect of reducing the opportunities for local applied research to adapt foreign technology, thus providing little incentive for local R & D and having a negative effect on links between R & D and marketing.

Restrictions on MNCs. These restrictions are almost universal. Many countries allow MNCs to operate in the country only if they establish joint ventures with local companies; some countries require majority local ownership of such ventures, or even as much as 60% local ownership. In some cases, certain industries are reserved completely for local ownership. There may also be restrictions on the amount or percentage of profits that can be repatriated by MNCs.

In order to operate within the constraints imposed by such restrictions, the links between R & D and marketing within a particular MNC may have to be replaced by market links between the company's R & D section and the local partner or distributor.

Property rights. The property rights to new technologies are strengthened through patents. Some developing countries have patent regulations but do not have the judicial or administrative machinery to enforce them. In several cases, new agricultural technology is specifically excluded from patent regulations. Legislation on plant variety protection, which is a property right similar to patents, exists in both the USA and Europe but, in the developing world, is found only in Argentina and Chile.

Patents are a formal linkage mechanism between R & D and marketing in that they require the disclosure of a new technology to the public. Once a new technology is made public, other companies may attempt to produce a similar product, thus increasing competition within the industry. Patents also provide the basis for market links between companies that develop new technologies and those that want to market these technologies. In the absence of patents, companies protect their new technology through trade secrets, which restricts the flow of technological information and may reduce their incentive to license the technology widely.

Price controls. Many developing countries impose price controls on agricultural inputs and outputs. These regulations reduce or increase the profits that companies engaged in R & D and marketing can expect to make, and thus influence the amount of R & D and marketing being undertaken in a particular country.

Tax incentives. Some developing countries have tax incentives to encourage R & D (for example, writing off research costs for corporate taxes, or reduced import duties on machinery or chemicals required for R & D). Tax incentives in India and the Philippines have induced some companies to organize their research activities into separate research institutions; this may create an additional barrier between researchers and marketing personnel, and thus reduce the effectiveness of the company's research and marketing activities.

Summary and Conclusions

The private sector is playing an increasingly important role in agricultural research and technology transfer in developing countries and, in general, spends more money on linking the two activities than is the case in the public sector. To further benefit farmers in developing countries, governments should not only implement policies that improve private sector links but should also draw on the lessons provided by the private sector to improve public sector links.

Policies to Improve Private Sector Links

Government policies should aim at ensuring that public sector research, extension and input-supply complement the role being played by the private sector. As discussed earlier, government policies and regulations influence not only the amount of private R & D and marketing being undertaken in a country but also the nature of the links, and the type of people who operate them, within the private sector and between the private and public sectors. In the case of some of these policies and regulations, it is difficult to ascertain what changes in them would lead to improved links. However, there are certain areas where changes could have a positive impact on links.

Unbiased and well-publicized government trials of privately developed inputs (which are relevant to farmers' needs) would encourage competition on the basis of technology rather than of advertising. Release of information from these trials would also accelerate the diffusion of new technology. Government extension recommendations and procurement policies based on such trials would encourage the private sector to set research priorities that reflect farmers' needs.

Extension could play a role in training dealers in basic agriculture, the potential of various inputs and management practices, and the safe handling of agricultural chemicals. It could also encourage the development of industrial associations which would improve communication within private companies and between these companies and their clients.

The existence of well-defined property rights to new technology would have a positive effect on links in so far as it would encourage more transfer of knowledge by developers of new products. In some cases, if a developing country reduced the restrictions on MNCs and on technological imports, this would improve links with other countries (both developing and developed) which are involved in producing technologies which are important to that country.

Another area of government activity which can play an important role in improving private sector links is in the provision of agricultural education facilities at local universities and technical colleges. If marketing personnel received better technical training, for example, they would assimilate specific technical knowledge more rapidly and with less input from the company, as well as being better equipped to communicate with scientists and to assess farmers' needs.

Lessons for Public Sector Links

Private companies are more efficient than public institutions in developing new products primarily because their marketing personnel play a greater role than their public sector counterparts in decisions as to what R & D projects should be funded and when they should be terminated. For example, most successful Indian seed companies were started by marketing personnel and continue to be controlled by them. In multinational chemical companies, marketing personnel are involved both in strategic planning to set the overall research priorities and in the decisions as to whether or not to end research projects.

All this is in contrast to the procedures which characterize public sector links. Here, decisions on research priorities are usually made against the background of little information on the potential market for research output. The result is that many projects continue through inertia to terminate them and/or develop products which are irrelevant to farmers' needs.

The Booz, Allen and Hamilton model suggests that efficient companies make a substantial investment in gathering and using marketing information in the early stages of the R & D process. Data gathered from a number of successful companies in developing countries, including BTC, Hindustan Lever Ltd and Mexican and Indian seed companies, indicate that they follow this model. Public research programs, on the other hand, rarely have the expertise to make effective use of market information in planning and managing the research process.

Most public agricultural research systems could increase their efficiency by investing more in social science research to assist in research planning and by improving the links between extension, government social scientists and private sector marketing personnel during the early stages of the research process. Extension agents, for example, usually have little "veto power" in decisions regarding the nature and duration of research projects.

Increased public sector efficiency would result in a reduction in the number of improved animal or crop varieties that farmers do not use and in the number of farm management recommendations that farmers do not adopt. However, bearing in mind the fact that a large proportion of privately developed technologies are not commercially successful (for example, 33% of new products developed and marketed by the private sector in the USA fail after they have been introduced into the market), even the most efficient public sector research programs will still produce some results that have little impact.

There are a number of measures which could be adopted to improve the flow of information within the public sector about new technology. If extension agents were involved in evaluating research projects, they would learn about the characteristics of a new technology as it progressed through the screening and field testing stages; this information could then be passed on to other extension agents. Another measure would be to include researchers in extension teams engaged in popularizing major new technologies, such as a high-yielding hybrid seed or a new fertilizer. To improve the flow of information in both directions, governments can increase the amount of personal contact between researchers and extension agents and dismantle as many status and institutional barriers as possible; in some of the larger public agricultural research systems, personnel could be rotated between research and extension.

Private companies have profit as their goal, and scientists and marketing personnel are rewarded with an increase in salaries when they contribute to an increase in profits. Communication between research and extension in the public sector would improve if the two activities had a common goal and an institutional structure that rewarded contributions towards that goal. Too often, public research and extension are separated into different institutions with different goals; researchers want to advance science, while extension agents want to spread technology. In addition, unlike the private sector, the cost of unsuccessful attempts to commercialize a new technology is borne by extension or public input-supply companies; researchers do not suffer any immediate consequences from developing technology which is not accepted by farmers.

The flow of technical information within the public sector may also benefit from more investment in modern communication devices, such as computers and electronic communications networks, and in training personnel to use these devices effectively. This would be particularly beneficial in large countries where public sector personnel are widely dispersed throughout the country. In many developing countries, large companies are making far more effective use of modern communication systems than public sector institutions.

Acknowledgments

We wish to thank David Kaimowitz, Hunt Hobbs and Gabrielle Persley for their comments on earlier drafts of the paper. However, the views expressed in the paper are those of the authors.

References

Booz, Allen and Hamilton, Inc. "A program for new product evolution." *Corporate Srategy and Product Innovation*. Rothberg, R. (ed). New York: Free Press, 1987.

Crosby, E. "A survey of US agricultural research by private industry." *Agricultural Research Policy Seminar 1986*. Hoefer, F., Pray, C.E., Ruttan, V.W., and Echeverría, R.G. (eds). Minneapolis: University of Minnesota, 1986.

de Andrade Alves, E. "Notes on dissemination of new technology." *Brazilian Agriculture and Agricultural Research*. Yeganiantz, J.L. (ed). Brasilia: EMBRAPA, 1984.

de Obschatko, E.S. and Piñeiro, M. "Política tecnológica agropecuaria y desarrollo del sector privado: El caso de la region Pampeana Argentina." The Hague: IS-NAR, 1985.

Echeverría, R.G. "Public and private sector investments in maize research. The case of Mexico and Guatemala." CIMMYT Economics Program Working Paper. Mexico D.F: CIMMYT, 1989.

Eicher, C.K. "International technology transfer and the African farmer: Theory and practice." Department of Land Management Working Paper. Harare: University of Zimbabwe, 1984.

Hobbs, H., and Taylor, T.A. "Agricultural research in the private sector in Africa: The case of Kenya." Working Paper No 8. The Hague: ISNAR, 1987.

Mikkelsen, K.W. "Inventive activity in Philippine industry." Ph.D thesis. New Haven, Conneticut: Yale University, 1984.

Pray, C.E. "Private sector research and technology transfer in Asian agriculture: Report on Phase I." Economic Development Center Bulletin. Minneapolis: University of Minnesota, 1985.

Pray, C., Ribeiro, S. , Mueller, R., and Rao, P. "Private research and public benefit: The private seed industry for sorghum and pearl millet in India." Economics Group Progress Report. Resource Management Program. Hyderabad: ICRISAT, 1989.

Ruttan, V.W., and Pray, C.E. *Policy for Agricultural Research*. Boulder: Westview Press, 1987.

7

A Conceptual Framework for Studying the Links between Agricultural Research and Technology Transfer in Developing Countries

David Kaimowitz, Monteze Snyder and Paul Engel

Many studies and program evaluations have identified weaknesses in the links between institutions responsible for agricultural research and those concerned with transferring technology to farmers as a major obstacle to the development and application of beneficial new technologies in developing countries (World Bank, 1985). In response to this, the leaders of these institutions, as well as those who fund and oversee them, have attempted to identify policies and organizational structures that would strengthen the relationship between research and technology transfer.

A number of models have been put forward as possible solutions. Among the most prominent are the US Land Grant model, which combines research, extension and education in one institution; the Training and Visit system (T and V), which involves subject-matter specialists and regular training of extension workers; and farming systems research (FSR), which emphasizes the role of constraint diagnosis and on-farm trials. Other suggestions include setting up joint committees of various sorts and establishing or strengthening agricultural information departments.

Experience has shown, however, that it is impossible to come up with a set of general recommendations which would be appropriate in all circumstances. Solutions which work well in one context perform poorly in others. While some characteristics are common to all situations where technologies are successfully developed and delivered, these tend to be of a general nature; the specific mechanisms for maintaining the links between research and technology transfer vary considerably from one situation to another.

However, when asked for advice on how to improve the links, we should be in a position to say something more than "it depends on the circumstances." This paper presents a conceptual framework and a set of hypotheses which may enable us to offer more meaningful advice once our study has been completed. It does not attempt to prescribe solutions to the problems of linking research with technology transfer, although we have fleshed out our conceptual framework with relevant observations wherever we have felt able to do so at this stage in our study.

In particular, the paper addresses four basic questions:

- What linkage mechanisms exist and what are their characteristics?
- What contextual factors influence which linkage mechanisms are appropriate to use and how?
- Which of these contextual factors can be controlled or influenced by policy makers and leaders of research and technology transfer institutions?
- What limitations do contextual factors impose upon the use of linkage mechanisms?

The term "linkage mechanisms" refers to the specific organizational procedures used to maintain research-technology transfer links. "Contextual factors" includes all the factors that affect the use and relevance of linkage mechanisms. Some contextual factors are internal in that they can be controlled or influenced by the leaders of the institutions; others are external and are influenced by the institutions' broader physical, political and socio-economic environment (Merrill-Sands and McAllistair, 1988).

Contextual factors can be divided into political, technical and organizational factors (Lane et al., 1981). "Political" does not refer here to party politics or broad government policies but to institutional politics and the interest groups which play a role in them; among these groups are those which are internal (such as research and technology transfer personnel), those which are external (such as national policy makers, foreign agencies and private companies) and those whose involvement can be both internal and external (such as farmers). We need to know what role these groups play in the creation of values, rewards and sanctions which inhibit or facilitate collaboration between research and technology transfer institutions.

The technical factors are the activities and methods which are associated specifically with the development and transfer of different types of agricultural technology to different environments and target groups. The organizational factors include the division of tasks, resources and authority between different organizations and individuals, and the internal management and informal dynamics of each organization and its components.

In some situations, the research-technology transfer relationship is not the critical constraint, such that manipulating linkage mechanisms and the contextual factors that condition them would make little difference. Changes in other areas must come first. In those situations where the relationship is critical, the linkage mechanisms and contextual factors which can be manipulated and those which are fixed may vary in each situation. Management must, in each case, identify the factors that can be controlled, determine the options available, and make hard decisions.

This framework and the overall study of which it forms a part are intended to provide a road map for that process. It should help leaders of research systems find out what paths exist and where they lead. The specific routes to guaranteed improved performance of research systems are not yet known, but this paper gives some indications as to their general direction.

The paper opens with an elaboration of the key concepts of the framework, and then discusses the criteria for evaluating performance. This is followed by analyses of the political, technical and organizational factors which affect linkage mechanisms in the development and transfer of agricultural technology.

Key Concepts

Research and Technology Transfer

The terms "research" and "technology transfer" have both functional and institutional meanings. The functional meaning relates to the *tasks* involved in the development and delivery of new technology. The institutional meaning relates to the *institutions and personnel* responsible for carrying out this process. Throughout this paper we have used these terms in both senses, as is common practice; it will be evident from the context in which the terms appear which usage is being referred to.

The main tasks of research are:

- discovery
- exploratory development
- technology consolidation

Discovery is the process of collecting information and/or searching for relationships between variables, the specific usefulness of which is as yet undetermined. This process is often also referred to as "basic research".

Exploratory development is concerned with the identification, understanding and control of the interaction between a proposed technology and the physical, economic and/or social environment in which this technology will ultimately be used. This process is often labelled "applied research".

Technology consolidation is the process of translating the results of basic and applied research into specifications for a new technology and ensuring that these specifications suit the type of farmers for whom the technology is intended. This involves some adaptive research, but it also includes all the work carried out to determine how to present and package a new technology and to identify exactly who might be interested in using it.

The main tasks of technology transfer are:

- technology production
- delivery of technologies to farmers
- monitoring and evaluating the use of technologies

Technology production is the process of producing the materials (physical inputs and/or information) in sufficient quantity and of making these materials available to those responsible for technology delivery.

Technology delivery is the process in which the technology is promoted and distributed to farmers. In most instances, agricultural technology is delivered through many channels and over varying lengths of time; as a result, what the farmers receive is often incomplete and contradictory.

Monitoring and evaluating the use of technologies involves ascertaining whether farmers have acquired the new technology, assessing the extent to which they adopt, adapt or reject it, and identifying the reasons underlying their response to it.

Implicit in the tasks outlined above is the assumption that they occur in a logical sequence; indeed, common sense and much of the available literature support this assumption (McDermott, 1987). In practice, however, many of these tasks may be performed simultaneously. Work may begin with exploratory development rather than discovery, or new research may be carried out on a technology already in the process of consolidation.

A variety of institutions and personnel play a part in carrying out research and technology transfer tasks. It is also important to note that many research institutions and personnel may be involved in producing, delivering and evaluating new technologies, while many technology transfer institutions and personnel may be active in discovering, developing and consolidating new technologies.

Technology Transfer or Extension?

We have used "technology transfer", rather than the more familiar term "extension", throughout this paper, apart from a few contexts in which national extension services are specifically discussed. The reasons for this decision are:

1. It is important to include the role of inputs and services in the discussion of technology development and delivery. This broader view is captured by the term "technology transfer", whereas "extension" implies a more limited focus on education/information.

2. Some activities associated with "extension", such as informal education in nutrition and health, are not within the scope of this paper.

3. "Extension" is now usually associated with conventional public sector extension services. "Technology transfer", however, can be applied not only to these services but also to those provided by many other institutions or organizations, such as private firms, parastatals, non-governmental organizations, formal educational institutions and producers' associations.

In this paper, "technology transfer" is not restricted to meaning a one-way flow of materials and information from those who develop and deliver the new technology (usually professional and paraprofessional personnel) to those who use it (the farmers, who are often mistakenly assumed to be less knowledgeable than the first group). "Technology transfer" implies a two-way flow of technical information between these groups. Materials and information are never simply "transferred" to the farmers; they are adapted and assimilated. Farmers do not only receive materials and information; they also provide information, both to other farmers and to those who are responsible for delivering materials and information.

Institutional Agricultural Technology Systems

An agricultural technology system (ATS) consists of all the individuals, groups, organizations and institutions engaged in developing and delivering new or existing technology. This definition is somewhat different from that of Röling (1988) and others, in that we make no assumption that the different institutions in the system work together or in a compatible fashion, nor are we using the word "system" in the dynamic sense commonly found elsewhere in the literature. ATS participants may nonetheless be linked in terms of their geographical focus or in terms of their focus on a particular commodity, or both (Engel, 1988). "New technology" refers not only to technology that has been recently developed but also to older technology which is being introduced to a new area or a new group of users.

In many ATS, some sources of information, knowledge, physical inputs and services may be entirely unconnected with any formal institution, but this feature is not within the scope of this paper. We are concerned here only with those parts of an ATS in which a set of formal institutions or units is involved; to denote this, we have used the phrase "institutional agricultural technology system(s)" (IATS) in this paper.

In order to carry out their various research and technology transfer tasks, IATS engage in a number of basic activities. These activities can be categorized into:

- those concerned with problem identification and with the acquisition, transformation, storage, retrieval, dissemination and use of knowledge
- those concerned with the production of material goods, including conceptualization, design, prototype production, testing, multiplication, packaging and distribution
- those concerned with the management of and administrative support for the above activities

In all three categories, there are various types of skills involved; these range from specific technical and socio-economic skills to more general managerial, communications and participation skills. This variety of skills, combined with the fact that most IATS encompass many different client groups, agro-ecological and administrative regions, products, approaches and disciplinary fields of interest, makes even the smallest IATS quite complex.

Links and Linkage Mechanisms

As indicated above, "research" and "technology transfer" have both a functional and an institutional meaning. Thus, the links between them may be discussed from two points of view: they may be seen as functional links, which relate to research and technology transfer activities; or as institutional links, which relate to the institutions and personnel that carry out these activities. In the former case we are thinking of links as activities which aim to form a bridge between research and technology transfer. In the latter, we are discussing the exchange of resources (such as information, money, labor and materials) between institutions and personnel. In this paper, the general term "link" is usually used, since both viewpoints are normally included in the discussion. However, there are a few contexts in which we specify our viewpoint by using the terms "functional links" and "institutional links".

The organizational procedures used to establish, maintain or improve links are termed "linkage mechanisms". These mechanisms can be characterized according to the following attributes:

- whether they are formal or informal, regular or ad hoc, mandated or voluntary, permanent or temporary
- whether they are facilitative mechanisms (that is, they provide resources) or control mechanisms (that is, they determine how resources should be used) (Leonard, 1982a)
- the amount and type of resources exchanged
- the administrative level at which they operate
- whether they focus on programming activities or are concerned with implementation or evaluation
- the numbers of individuals involved

A scale can be created going from the least to the most demanding types of linkage mechanisms. Mechanisms for facilitating the exchange of information would be at the lower end of this scale; those for implementing joint activities would be at the higher end; and those for the joint planning of independently implemented activities would lie somewhere in between.

Formal and informal links. The degree to which a link is formal refers to whether or not it is given official sanction (Snyder, 1988). In theory, formal linkage mechanisms follow officially specified patterns, whereas informal ones do not, being built on personal relations. In practice, the distinction between the two is less clear cut: most formal interactions have informal aspects, and vice versa.

Formal linkage mechanisms mentioned in the literature include: committees, task forces, liaison units and officers, agricultural communications units, pre-extension units, subject-matter specialists, joint activities, contracting research by development agencies, farming systems programs, publications, presentations and demonstrations, staff exchanges, inter-agency agreements, service provision, matrix management, joint plans, shared supervisors, policy mandates, and meetings.

Informal mechanisms consist of the exchange of resources and information without official sanction or through personal contacts. Communications studies have found that people who maintain personal contacts outside their unit play a key role in inter-unit communication.

Just because a mechanism is informal does not mean it cannot be managed. Management can either foster or hinder the establishment of informal links. This can be done by changing the physical proximity of groups, promoting joint social activities, encouraging staff rotations, publicly sanctioning informal contacts, placing people in certain positions on the basis of their compatibility and previous personal ties, and a number of other measures.

Institutionalization refers to the degree to which a pattern becomes routine and follows set rules. For the most part, institutionalized mechanisms are more permanent and formalized, ad hoc mechanisms more temporary. Ad hoc and temporary mechanisms, such as task forces, have the advantage of being designed to meet a specific objective. Their extraordinary nature can create a sense of urgency. Institutionalized mechanisms permit the development of mutual expectations and can be improved over time. Although there are important exceptions, recurrent problems lend themselves more to formal approaches.

Criteria for Evaluating Links

Any discussion on improving the relationship between research and technology transfer requires some idea of what constitutes a good relationship. We have established five criteria for evaluating the links which form the basis of this relationship:

- IATS integration
- availability of new technologies
- relevance of new technologies
- responsiveness of new technologies to needs of resource-poor farmers
- institutional sustainability

These criteria will enable us to study links from a purely analytical and objective standpoint. Although only one criterion, integration, refers to the links themselves, high performance on the others provides indirect evidence of effective links.

The criteria chosen here are not necessarily those which are used in IATS to evaluate links, for such criteria often contain a more subjective element in that they reflect not only the official goals of an institution but also the goals of the individuals within it. This is an important point, because too often people who evaluate IATS assume that the individual goals are the same as the official goals; in reality, however, each individual has his/her own set of personal, institutional, social welfare and/or political goals over and above the official goals. These individual goals may be both rational and legitimate, and they should be taken into consideration when seeking to understand the behavior of an institution and the staff within it, but they do not provide a basis upon which to evaluate the efficiency and effectiveness of an IATS.

All IATS will perform better with regard to some criteria than to others. Although we have used several criteria, no attempt has been made to weight them; each one is regarded as just as important as the others and must be examined independently. Neither has any attempt been made to produce an overall success indicator. Instead, the criteria are best used simply as a checklist. Policy makers and managers may find there are trade-offs in their achievements.

The criteria are defined here; a brief note is added on the impact of new technologies on welfare. We will then examine how political, technical and organizational factors affect the performance of IATS in relation to these criteria.

Definition of the Criteria

IATS integration. The idea that a high level of coordination, collaboration and communication within an IATS is a prerequisite for high system performance constitutes our first criterion, integration. The level of integration is gauged according to the amount of resources exchanged between the parts of an IATS and the importance that each part attaches to these resources.

IATS which regularly make available relevant new technologies exhibit high levels of integration between research and technology transfer.

However, it must be pointed out that the existence of a high level of integration in an IATS is not a guarantee that relevant new agricultural technologies will regularly be made available, because there are other conditions that must also be met. There is little value in coordination, collaboration and communication for their own sake. Similarly, although a low level of integration will contribute to the failure of an IATS to regularly make new technologies available, it need not necessarily be the only reason for this failure.

High levels of integration do not necessarily imply the absence of conflict between researchers and technology transfer workers. And where conflict exists, it may make a more positive contribution to the research-technology transfer links than is often thought; it can prevent stagnation, highlight important issues which might otherwise be overlooked, stimulate both groups to work harder, foster creativity and provide a forum for problem solving (Arnold and Feldman, 1986). High levels of integration are best achieved by effectively managing conflicts, not trying to suppress them.

Another important aspect of integration is efficiency. Integration is costly in terms of time, money and other resources and generally involves a reduction in autonomy. Some integration is necessary, but beyond a certain point devoting additional resources to integration in preference to other

Figure 1. The relationships between the performance evaluation criteria

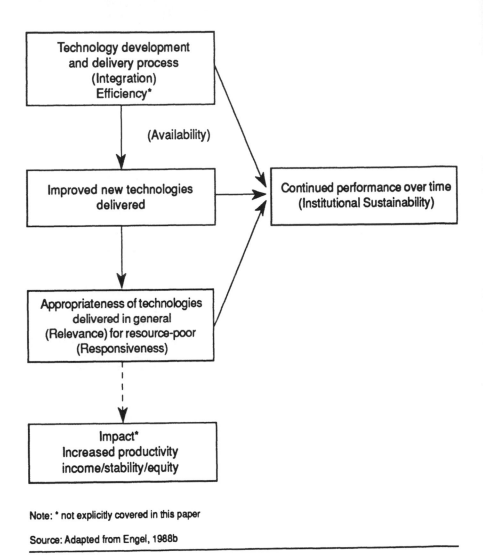

Note: * not explicitly covered in this paper

Source: Adapted from Engel, 1988b

activities will be counterproductive. On purely theoretical grounds, efficiency is important enough to be included as a separate criterion. In practice, however, it is difficult to assess.

Availability of new technologies. The term "availability" is used to cover the process in which a new technology is invented, technology transfer workers and producers are made aware of it, and producers are provided with access to the inputs and services necessary to use the new technology. The degree of availability depends on how much technology is produced, how effectively it is promoted and how reliable and convenient the inputs and services needed by the producers are.

Relevance of new technologies. A new technology is relevant to a group of farmers if it responds to their needs. The best way to assess the relevance of a technology that is widely available is to look at the extent of its adoption. There are a number of variables which determine the extent of adoption, including the profitability and social acceptability of the technology, its importance to producers' systems of production, and whether or not it was developed in response to a clearly articulated demand from producers or technology transfer workers.

Responsiveness of new technologies to the needs of resource-poor farmers. This is exactly the same criterion as relevance, only it refers to whether the technologies are relevant for resource-poor producers as opposed to other types of producers. It has been included because there is strong evidence that the linkage mechanisms required to serve high-resource farmers may be significantly different and, generally, easier to develop than those needed to serve their poorer counterparts.

Institutional sustainability. IATS which seem quite successful according to an evaluation carried out over a particular period may seem less successful in an evaluation carried out at a later date. Thus an important evaluation criterion is the ability of an institution to sustain its performance. Because of the sustainability issue and the fact that technology development and delivery is a slow and often discontinuous process which may take years before coming to fruition, performance should be assessed only over a long period.

Impact of New Technology on Welfare

None of the criteria which have been outlined above relates specifically to the impact of IATS on the welfare of producers and consumers.

The first criterion, integration, deals only with the system's efforts, not with the results of those efforts. Availability focuses on the ability to produce and deliver outputs, not on the impact those outputs may have (Snyder, 1988). Sustainability deals with the performance of institutions, not with the impact that that performance has on producers. Only relevance and responsiveness to the needs of resource-poor producers are in some sense connected to impact in that producers would be unlikely to adopt new agricultural technologies which adversely affected their welfare.

The impact of new technology on welfare could be examined on the basis of such aspects as increased farm income, reduced risk, resource conservation, improved health, better security and overall economic growth. However, because so many variables affect these aspects, it is almost impossible to establish a direct correlation between research-technology transfer links and the impact of new technology on welfare.

Political Factors

The political factors which influence research-technology transfer links can be divided broadly into:

- those which determine the external and internal pressures on institutions and personnel within IATS to achieve high levels of performance
- those which determine the quantity and quality of the resources within IATS

With regard to the first group, imposing goals on institutions and personnel within IATS and providing them with the resources needed to meet these goals would have little effect unless they had the *desire* to achieve such goals; that desire depends largely on what incentives are provided. Regarding the second group, many political factors come into play in determining whether or not IATS have adequate resources to fulfil their mandates.

In essence, then, without adequate resources *and* the desire to use them effectively, no mechanism created to improve links is likely to produce satisfactory results.

Political Pressures

In the absence of positive external pressure from national policy makers, foreign agencies, farmers or the private sector, the dynamics of an institution tend to be dictated by internal processes, resulting in poor performance.

This hypothesis is borne out by the situation found in many low-income countries and is the result of historical factors and current political and social structures (Sims and Leonard, 1989).

The historical legacy. In most low-income countries there is a marked difference between the historical legacy of the commercial agricultural sector and that of the subsistence sector. In the case of the former, foreign settlers and indigenous landed elites had close and generally direct contact with researchers. Researchers endeavored to meet commercial farmers' needs for several reasons: they felt obliged to do so; they had a relatively small group to deal with; and they had similar backgrounds to the farmers and therefore could communicate relatively easily.

Responsiveness to the needs of the subsistence sector, however, was poor. Little or no technology was produced specifically for this sector, and where extension services existed they often focused on non-technological activities. Research-technology transfer links were characterized by a great difference in status between researchers and technology transfer workers and between the latter and farmers. Information flowed only one way, from the researchers "down" to the farmers. Responsiveness to the needs of resource-poor farmers would often increase significantly in times of famine and other similar crises, only to decrease again once the crisis was over.

Thus, in terms of the criteria defined above, the commercial sector benefited from substantial technology availability and relevance, as well as from institutional sustainability; there was integration in a sense, but much of it was directly between researchers and producers. For the subsistence sector, however, IATS performed poorly in relation to all the criteria.

Current political and social structure. Where external pressures on IATS have not intervened to change the historical pattern, it has persisted or, in some cases, degenerated to the point where the attempts being made to meet the needs of either the commercial or the subsistence sector meet with little success.

In many cases, the IATS in low-income countries face little external pressure to improve the links between research and technology transfer other than that applied by foreign donors. For political reasons, governments are reluctant to allow farmers' organizations to be formed or to become too powerful. Usually, the only types of organizations which are found among resource-poor farmers are informal networks of the patron-client type; stronger members take on the role of patrons, while the rest of the members assume the role of clients. The members of these networks exchange goods and services, with most of the benefit accruing to the patrons. They may take advantage of the credit, inputs or services which are offered by IATS but they rarely exert pressure on the IATS to produce new technologies.

The dominance of patron-client politics in low-income countries has a twofold effect on research-technology transfer links:

- technology transfer services come under pressure to provide more than just advice and are pushed towards concentrating on activities likely to make them less integrated with research; for example, they offer credit and inputs rather than information, or concentrate on servicing the needs of the patrons, who may have some political power, rather than those of the clients, who have none

- research institutions tend to become oriented towards the rest of the scientific community or towards their hierarchical superiors, which results in the tendency among researchers to prefer to do on-station rather than on-farm work, to concentrate on export crops and to live in urban areas where they can interact with people of similar backgrounds rather than in remote areas where they would have more interaction with resource-poor farmers

In essence, the lack of external pressures may result in institutions and personnel becoming motivated more by their own social and political needs than by the needs of resource-poor farmers. Many of these institutions suffer from lack of funding, and this further reduces the level and quality of the work done by their personnel.

Effects of external pressures. As implied in the hypothesis, good institutional performance requires positive external pressures on IATS by national policy makers, foreign donors, farmers and the private sector. The nature of these pressures is described here; this is followed by an outline of how external pressure may, in some cases, adversely affect research-technology transfer links.

National policy makers. Generally, national policy makers intervene forcefully in technology issues only in exceptional circumstances, such as disease outbreaks, major crop shortfalls, rapidly rising food imports, rural unrest, a highly publicized international breakthrough in technology or a radical change in government. At such times, they will exert pressure on IATS to cut through red tape and bottlenecks to produce quick results; new resources are brought in, objectives are clarified and there is an overall, albeit often short-term, dramatic improvement in performance. More consistent pressures are exerted by national policy makers usually only in those countries where one or a few crops play a dominant role in society, as in the case of rice in Asia or sugar in the Caribbean.

Foreign donors. This term includes multilateral and bilateral aid agencies, externally sponsored non-governmental organizations (NGOs) and international agricultural research centers (IARCs). These groups provide a substantial proportion of the resources required by national institutions in low-income countries; their tendency to concentrate their funding on program expenses, equipment and training, rather than on salaries, and their ability to elicit government matching funds for their projects give them greater leverage than their overall budget share might warrant.

Until recently, foreign aid agencies regarded research and technology transfer as separate systems. This approach is now changing, and they are making the improvement of links between the two systems a precondition for further funding. In addition, they are increasing financial support for the development of technologies relevant to the needs of resource-poor producers. Externally sponsored NGOs are carrying out innovative participatory projects. IARCs are providing researchers and, to a lesser extent, technology transfer workers with incentives to engage in more relevant work; they are also trying to mobilize more external funding for linkage activities, such as extension training, agricultural communication and liaison, farming systems research, social science programs and the use of subject-matter specialists. This has provided an incentive for those working in the national institutions to focus more attention on the links between research and technology transfer.

Farmers. As indicated above, resource-poor farmers in low-income countries are seldom able to exert pressure on national institutions, but there are situations in which they may benefit from the pressures exerted by other producers. This is most likely to occur where there is a group of more affluent and politically influential farmers who have the resources and incentives to invest in research-generated technologies. Although this may bias researchers and technology transfer workers towards the needs of this more affluent group and thus detract from efforts to meet the needs of resource-poor farmers, to the extent that the two groups of farmers grow similar crops, contend with similar agro-ecological conditions and face similar price structures and resource scarcities "resource-poor farmers may gain considerably more benefit from the political ability of the large owners to lobby for agricultural interests than they lose in the bias of system against their particular needs" (Sims and Leonard, 1989).

Private sector. Private companies influence public sector performance both directly and indirectly. Examples of direct influence are representation on public advisory boards, funding of public research projects, direct contact with researchers and technology transfer workers, and private (or public) delivery of publicly (or privately) developed technologies. An example of indirect influence is the implicit competition which takes place when private and public sector agencies are simultaneously involved in similar activities (Israel, 1987). The degree of private sector involvement and of its influence on research-technology transfer links depends upon the level of a country's development and on government regulations and incentives.

Although the involvement of private companies may strengthen some links between research and technology transfer, it may also bias public research and technology transfer towards producing capital-intensive technologies which have little relevance to the needs of resource-poor farmers.

However, as in the case of the pressures exerted by more affluent farmers, the spin-offs for the poorer farmers may outweigh this disadvantage, at least in the short term; in the longer term, because of the profit motive, private company involvement may mean that little attention is paid to the effect of new technologies on the physical environment.

An important aspect of private sector pressures on IATS is the influence exerted by large plantations and processors, particularly those with monopoly power. These concerns are usually in a position to finance technological activities and to make full use of new technologies; sometimes they develop and deliver technologies themselves, sometimes they contract out these activities to the public sector or a private company. This will have a positive effect in terms of all the evaluation criteria except responsiveness to resource-poor farmers, few of whom grow crops for processing.

Limitations of external pressures. In many cases, external pressures are heavily resisted by the institutions within the IATS. This is partly because of people's natural tendency to resist any incursion on their autonomy, but there are a number of other more valid reasons for such resistance.

Firstly, those exerting pressure often do not adequately understand the problems they wish to see solved. Thus they may demand results which are not feasible or cost-effective, may overlook potential dangers or secondary effects, and may place undue emphasis on short-term problems and on the symptoms of problems rather than the underlying causes.

Secondly, the technology development process is often long term, whereas external pressures often emanate from transitory and unstable sources. For example, frequent changes in government result in changes in national priorities and policies; within the international donor community, topics and approaches go in and out of fashion. If institutions always respond to these fluctuating external pressures by changing their structures and activities, the chances of building up the effective relationships needed to create sound research-technology transfer links are severely reduced.

Thirdly, competing external demands may have a very damaging effect on institutions. The emphasis placed on one aspect of an institution's activities by a foreign donor might conflict with the demands made by government ministries, and this conflict will be echoed in the institution's performance. In some countries, competition between donors has brought national institutions to a state of complete paralysis.

Fourthly, as already noted, external pressures often reduce rather than increase the responsiveness of researchers and technology transfer workers to the needs of resource-poor farmers.

Lastly, external pressures may force leaders of institutions to indulge in "window-dressing" in order to create the impression that they are responding to external demands. For example, if improvement in the links between research and technology transfer is a precondition for external financing, committees may be constituted and documents published to create the illusion this improvement is under way; but such manoeuvres may bear little relation to the real situation (Röling, 1989). Although window-dressing may have some positive results, it does add to the workload of institutions and it makes critical assessment of linkage mechanisms difficult.

The Ability of IATS to Command Resources

The quantity and quality of resources available for technology development and delivery varies according to region, country, client group and commodity. In general, high levels of appropriate resources are associated with:

- agricultural products which are strategically important because they generate foreign exchange or are staples in the diet of the urban population
- client groups who have the ability and incentives to exert pressures on technological institutions
- favorable agro-ecological and socio-economic environments in which there is substantial use of purchased agricultural capital goods

A more tentative relationship exists between those IATS with high resource availability and the "size" of the commodity, client group or area they serve. Size is difficult to define; possible factors on which a definition could be based are value of output of the IATS' clients, the number of people served and the availability of resources which can be tapped to support technological activities.

Greater access to resources implies the ability to sustain larger, more sophisticated institutions. This assumption underlies the following hypothesis:

IATS which have high resource availability are more differentiated than those with low resource availability, leading to more complex, well-endowed and sophisticated linkage mechanisms.

IATS with high resource availability are generally characterized by a greater division of labor than that found in IATS with low levels of resources, and by a greater ability to make use of slack resources, to allocate more funds to linkage-related activities and to create more structured and formal linkage mechanisms (Stoop, 1988). Researchers and technology transfer workers in well-endowed IATS tend to be from similar backgrounds and to share similar values, which promotes better communication and empathy between them; however, it should be noted that this communication is likely to suffer if these personnel become too specialized.

High-resource IATS are generally those in areas which offer a relatively wide range of amenities (schools, hospitals, cultural opportunities, etc) for researchers and technology transfer workers. Hence, these IATS are able to recruit and retain more educated, specialized, higher caliber personnel, which in turn promotes more effective communication. As technology transfer workers become better educated, they are more able to assume responsibility for adaptive research and specialist tasks formerly handled by researchers.

Farmers served by high-resource IATS are often better educated and organized and thus more able to exert pressure on institutions and to understand the information provided by them. There tends to be more direct contact between farmers and researchers in these circumstances and a larger variety of channels through which farmers receive and provide information (Stoop, 1988).

In the light of the points outlined above, it might well be asked what can be done for those regions, countries, client groups and commodities where the quantity and quality of resources commanded by IATS are low. There are three possible courses of action.

Firstly, an attempt could be made to improve the resource base through the use of people who have an ideological commitment to working in situations where others, motivated solely by material considerations, would not be willing to work. Such people can be found within NGOs which have a humanitarian or religious base; other possible candidates are politicized professionals and, in developing countries ruled by highly ideological regimes, young adults.

Secondly, the tasks carried out by IATS can be simplified to allow them to be performed with the resources, particularly human resources, that are available. It may be feasible, for example, to carry out farmer-to-farmer interchanges, simple trials and practical experiments with new plant species using relatively limited local resources. More use can be made of paraprofessionals and farmers. Although the results of such efforts will probably be more limited than those when specialists are involved, some results are better than no results.

Thirdly, efforts can be made to provide disadvantaged groups with skills and levels of organization that will enable them to interact effectively with the institutions in the IATS and to demand resources from policy makers and external agencies. In some cases it may be more effective to devote any available resources to creating this organizational capacity than to spend them on the IATS.

In discussing the ability of IATS to command resources it is necessary to distinguish between resources which are externally generated and those which are generated from within the group or area the IATS serve. As noted previously, resources which are externally generated may be unstable.

Internally generated resources might be more stable in those situations where the relevant group's own resources and its concern with technological issues are relatively stable; groups which provide IATS with resources during a crisis, or are vulnerable to fluctuations in the prices of their products, are unlikely to be able to sustain their efforts in the long term.

Technical Factors

What type of research-technology transfer links are most appropriate depends a great deal on the nature of the activities the IATS is assigned to carry out. This section discusses the technical factors relevant for linkage design. It looks first at the problem of how to involve the farmer in technology development and delivery and then examines how the activities associated with these tasks vary according to the types of environments and technologies involved.

Farmer Input and Targeting

Linkage mechanisms that give farmers and technology transfer workers opportunities for input and feedback early on in technology development and the accurate identification of target groups are both required for the production of relevant new technologies.

Links may be direct, consisting of participation by farmers in setting the research agenda or of diagnostic research in the farming community to assess user preferences and needs (Röling, 1989). Alternatively, inputs and feedback may be channeled through technology transfer workers, who then serve as an indirect link.

For these links to be effective, producers, researchers and technology transfer workers will often have to be taught participation skills to allow them to interact effectively with each other. These skills may include learning local languages, using instruments to obtain technical measurements, articulating needs and taking part in experiments.

The early targeting of user groups is a prerequisite for the successful development of new technology. Within the broad category of agricultural producers are many subgroups, each with its own technological requirements. These subgroups and their needs must be identified, and technology development and delivery must take their existence into account. This targeting process is closely related to what the farming systems literature calls "identifying recommendation domains" and commercial marketing research refers to as "market segmentation" (Röling, 1989).

Environmental Diversity

The level of integration and the complexity and/or differentiation of the tasks performed by IATS must increase as the environment becomes more diverse or unknown.

Complex tasks are those involving many variables, high levels of abstraction and sophisticated analysis. To carry out such tasks, institutions must have highly trained staff from a wide variety of backgrounds. Often, complexity is also associated with the dispersion of work locations (Snyder, 1988). To be handled effectively, complex tasks require a more open communication system than that found in hierarchical decision-making structures, and flexibility at lower levels in determining appropriate technological responses (Lane et al., 1981). Decentralization of authority, whether formal or informal, is also essential (Martinez, 1989).

Hierarchical systems are those with heavy constraints on communications outside the vertical authority channels, more authoritative decision making and greater status differentials (Lane et al., 1981). Examples are the T and V system of extension, agricultural technology promotion campaigns, and commodity systems such as the Kenya Tea Development Authority (Chambers, 1988). Such systems are normally successful only where few commodities are grown in relatively uniform, controllable and predictable conditions.

Thus, task complexity is closely related to environmental diversity. This is especially marked outside the fairly uniform green revolution areas. Physical and biological diversity is found in arid areas but is most pronounced in semi-arid, subhumid and humid zones. Physical variations within the same field can require different crop varieties or combinations. Soil, slope and vegetation differences compound the problem, while multiple canopies of plants, multiple tree-crop-livestock interactions and the sheer number of different species used can be bewildering. Moreover, social diversity is interwoven with environmental diversity, such that each place and social group can be seen as unique, requiring its own path for development (Chambers, 1988).

Diverse environments require more location-specific diagnosis of constraints and the adaptation of technologies. As a result, research efforts must be more widely dispersed. This dispersion, although it separates researchers from one another, often brings them into closer contact with technology transfer workers, offering opportunities for increased communication.

The most marginal farming systems tend to be the most complex and diverse and to face the greatest risks. Rainfed cropping systems in upland areas are generally both less productive and more diverse than irrigated systems. These environments pose more complex technical problems not only because of the multiple activities associated with them but also because less is known about them and the constraints are greater.

If IATS are to perform as well in these environments as they do in more homogenous ones that are better endowed, they must accomplish more complex tasks. This, in turn, requires features typically found only in well-endowed IATS. The more difficult environments are usually served by IATS with very limited resources. "There are far fewer scientists per farming system, both because of the scarcity of scientists and because of the many farming systems" (Chambers, 1988).

The adaptive, problem-solving approaches demanded by these diverse environments require levels of experience, education and professionalism that cannot usually be found among those working there at present. Most people with alternative employment opportunities prefer not to work in these environments, and leave after short periods of time. This imposes strong limitations on the levels of performance achievable. Thus, producers with the greatest need for a sophisticated IATS are least likely to have one.

Other Environmental Factors

Other important environmental factors which affect IATS' tasks include:

- the availability of communications channels and infrastructure
- the development of the necessary infrastructure and traditions for farmers to make use of inputs and information produced outside their communities
- the level of pre-existing knowledge about the environment
- the dispersion and accessibility of the farming population

The choice of communications channels that could be used as links will depend on producers' access to and ability to use them. Thus the level of literacy among producers, the availability of television, radio, telephones and reading materials, and the way producers normally use these channels, have an important bearing. Where input distribution channels, particularly those in the private sector, are weak, extension services often concentrate on input delivery. Dissemination of technical information becomes a less important part of their work, reducing the links with research.

Researchers face limitations in the types of technologies they can productively work on, since for many inputs the necessary infrastructure is simply not available to produce and distribute them. Furthermore, as we have already seen, producers who make little use of research-generated technologies, particularly purchased inputs, are less likely to exert pressure on their IATS for results.

Knowledge of the environmental conditions, farming systems and technologies that producers work with also has strong linkage implications. As recent literature has shown, producers have a considerable amount of practical knowledge to contribute regarding the regions, technological regimes and systems for organizing production with which they are familiar (Tripp, 1988). The same may also apply to technology transfer workers. However, this advantage disappears when these groups are faced with new situations, as is the case when farmers are resettled, radically change their farming system, or move from individual to collective production. In these unfamiliar circumstances, the input from producers and technology transfer workers may still be important, but it will reflect preliminary impressions rather than detailed knowledge.

Research, when faced with new environmental conditions, often has to concentrate on basic exploratory work, and in the short term has little of practical value to offer. When young institutions are pressured to produce quick results at the stage when they are still putting together the knowledge base to respond to their task, the results are often disastrous. Progressing prematurely to technology consolidation in these conditions may be especially dangerous. In these situations researchers and technology transfer workers have the greatest need for information from producers.

When services are provided to dispersed and inaccessible farm populations, researchers and technology transfer workers have fewer opportunities for direct interaction. To be effective, technology transfer workers must be close to the population they serve. Research, however, must for reasons of cost be concentrated in relatively few locations. The resulting lack of contact between the two groups is not necessarily bad; in many cases direct contact is not the most effective or efficient means of linking research and technology transfer.

The relative dispersion and inaccessibility of researchers and technology transfer workers increases the need to decentralize decision making on minor administrative matters. If such decentralization does not take place, communications problems between the central offices and the field locations can paralyze operations and/or make those activities which do occur less relevant to local conditions.

Activities Associated with Different Types of Technology

Different types of technology require different linkage mechanisms; one set of mechanisms will not be adequate for IATS which deal with a wide variety of technology types.

Discussions on how linkage mechanisms work tend to be based partly on unsubstantiated generalizations. Most existing literature implicitly takes as a model the links required to develop new plant varieties. There is little reason to believe that this pattern can be applied to other technologies.

Technologies should be classified into different types only if they require distinct links for their development and delivery. The broad types we have so far identified include:

- existing and new technologies
- physical inputs and information
- private and public goods
- complicated and simple technologies
- centrally and locally generated technologies
- producer-, research- and policy-driven technologies

Existing and new technologies. *Specific linkage mechanisms are required to effectively develop and deliver new technologies in addition to the mechanisms used for delivering already existing technologies.*

Most of this paper discusses the development and delivery of new technologies. Much of the work within IATS, however, involves technologies which are already well established, at least nationally or internationally, for which the IATS does no original research or adaptation.

Most links between research and technology transfer concern such already established technologies. For example, researchers often give extension workers lectures on the production of a specific crop based on the general state of the art rather than on new trial results or a new technology. Nor is any new technology involved when a technology transfer worker comes to a researcher with a sample from a diseased crop and asks for assistance in identifying the pest which caused the damage. Similar comments could be made regarding a wide variety of support activities which researchers typically provide to technology transfer workers, such as drafting manuals or recommendations, providing laboratory and library facilities, and backstopping extension activities.

In high-performance IATS, most researchers will play some, even if only a small, role in technology transfer, and most technology transfer workers will play some part in research. In addition, those who work on the exploratory development of a new technology should also be involved in its consolidation and production.

Product champions are essential for the development and delivery of new technologies. These are people who have both sufficient interest and authority to push the new technology through the development and delivery process and help to overcome obstacles (Peters and Waterman, 1982).

Work with already established technologies does not necessarily require either product champions or the direct involvement of researchers. In fact, most high-performing IATS shield researchers from having to devote a large proportion of their time to this type of work in order to ensure they have sufficient time for their primary responsibilities.

To produce and deliver new technologies requires substantial modifications in the technology transfer infrastructure (in the case of physical inputs). This slows down the rate at which they become available. More contact between research and technology transfer when and/or before the technology is being consolidated reduces this time lag (Snyder, 1987).

Crops and other technologies with which producers and technology transfer workers are completely unfamiliar have similar implications to those described for new environments. In other words, researchers, technology transfer workers and producers must work closely together to ensure that they gain maximum advantage from each other's insights.

Physical inputs and information. Some technologies take the form of physical goods. Others involve only information or cultural and management practices. The units which must be linked, the predominant communications channels, and the output control mechanisms required are different in the two cases.

The delivery of physical inputs requires a set of actors and roles which does not exist in the case of pure information technologies. These actors include input producers and distributors and, where high levels of investments are involved, credit agencies. The presence of these additional actors/roles greatly alters the linkage dynamic. Whereas educational materials, both scientific and popular, lie at the heart of links in the case of information technologies, product distribution and market promotion are more important when physical inputs are involved.

The relation between research and input suppliers provides a potentially important additional channel for user feedback and market information. Indeed, the importance of the links between research and the supply of inputs can eclipse extension's role in disseminating technical information. Thus, breeders' relations with seed multipliers can prove more important for transferring new varieties than their relations with extension.

A more formal process for approving recommendations is generally advisable for physical inputs, because it is more costly to produce or import a new item than to recommend a new cultural practice. New products may also pose higher health, safety or environmental risks. In the case of new plant material, seed committees meet to decide whether a new variety should be released. Formal requirements are usually established for determining a pesticide's effectiveness and toxicity before it can be sold. These processes provide a forum for interaction between researchers and technology transfer workers. In contrast, a new recommendation for planting dates, pruning methods or similar practices need not be subject to a formal review process.

Private and public goods. High performance according to all our criteria, except responsiveness to the needs of resource-poor farmers, is more likely if the technologies are private goods. Most physical inputs are private goods, the main exceptions being goods which can be produced easily by farmers, such as self-pollinating plant varieties and natural fertilizers. As these inputs are less profitable to produce and farmers may have no interest in purchasing them, it is often as difficult to achieve high performance with them as it is with pure information technologies.

Complicated and simple technologies. Technologies which are more complicated to use or produce require greater and more sophisticated educational efforts. Manuals and/or intensive training efforts may be required. Researchers will probably need to be in regular direct contact with manufacturers.

Skills training for producers and even for technology transfer workers is often required for using complicated technologies. This, in turn, requires changes in the roles of researchers, technology transfer workers and producers. If use becomes very complicated, specialists (veterinarians, professional fumigators, tractor mechanics, etc) may replace farmers as the main users. The use of these complicated technologies in concentrated areas (such as large irrigation projects, capital-intensive horticultural concerns or fully mechanized farms) lends itself particularly well to the development of these specialized groups.

The livelihood of these groups depends on detailed knowledge of the research-generated technologies with which they work. This makes their interaction with researchers and technology transfer workers quite different from that of most producers in developing countries at present. They have more direct contact and make greater use of specialized communications channels. In time, farmers too may become more sophisticated in their approach to research-generated technologies, as their enterprises become more specialized.

Centrally and locally generated technologies. Certain technologies lend themselves to being generated in one or a few central locations. Others do not.

Technologies applicable over wider areas or in many situations can more easily be generated from central locations. For example, a new pesticide may be developed at the international headquarters of a multinational corporation for use around the world. Other technologies have only very local applicability and require multilocational field trials or other adaptive research activities.

Research on topics such as livestock and perennial crops tends to be concentrated in a few places because it is both costly and complicated.

While the need for adaptive research may be great, such trials are expensive. Thus only a small number of trials can be done and the potential losses caused by doing them badly are very high. This research is also longer term and more difficult to do on farm (data requirements are heavy and farmers are less willing to risk their animals or tree crops).

Economies of scale in the production of inputs favor the concentration of research. Even if it is preferable to have a wide variety of pesticides, inorganic fertilizer formulae or tractor models to meet local conditions, producing them is usually prohibitively expensive. (Economies of scale also affect the organization of input delivery, and thus have other linkage implications).

When research is not concentrated, the physical dispersion of researchers makes them more directly accessible to technology transfer workers, whose knowledge of local conditions is likely to be relevant for the generation of location-specific technologies. Technology transfer workers also have more opportunities to become involved in research when this consists of a considerable number of decentralized, low-cost field trials with relatively unsophisticated data requirements.

Producer-, research- and policy-driven technologies. Technology transfer workers and producers concentrate their demands for research on the problems which they perceive as urgent. Frequently these concern pests or diseases. These groups also pay more attention to technologies that offer a clear short-term advantage than they do to those that appear only marginally superior to current practices or that require effective management to bring substantial benefits.

When clearly advantageous technologies become available, a considerable amount of pressure may be exerted on research for additional information and adaptation. Most research, however, tends to concentrate on the less spectacular technologies or on providing maintenance to sustain technologies which have been developed previously. The incremental improvements thus provided are harder to perceive, and hence they elicit less interest and participation.

Producers and technology transfer workers rarely emphasize long-term or less obvious problems such as preventative (as opposed to curative) health issues or resource conservation. Röling refers to technologies responding to these latter problems as "policy-driven", because getting producers to adopt them usually requires incentives provided by policy makers (Röling, 1989). These incentives can be positive (bonuses, subsidized credits and inputs) or negative (regulations, sanctions) and must be incorporated into the overall activities of the IATS.

As a result of the bias in the type of technologies demanded by producers and technology transfer workers, some researchers will be under constant pressure, while others will be practically ignored. Since performance improves when external pressure is high and there is producer input, performance for producer-driven technologies will tend to be better than for those technologies which are policy- and researcher-driven.

Organizational Factors

Institutional Structure

The range of tasks which are performed by IATS can be divided among institutions, units and individuals in a variety of ways.

Formal boundaries between different entities simultaneously increase the interaction of those within the boundary and limit access to those outside. They permit each entity to specialize in terms of the tasks it undertakes, the inputs it uses, the outputs it produces and the groups with which it interacts. While conflicts and diverging interests or strategies do not disappear, within the boundaries it becomes easier to accommodate them.

The evolution of institutional structures is a complex process. IATS change slowly through the interplay between competing interests. Personalities and informal links play an important role. Key decisions are made at many different locations within the government hierarchy. In addition, private companies, NGOs, producers' associations and other external agencies over whom the government has only a limited amount of control are now beginning to play a more important role in IATS. Differences in current structural arrangements can often be traced back to models copied from or promoted by different external groups.

While managers have some opportunities to manipulate structure, they frequently find themselves constrained by inertia, political opposition and existing legislation and regulations. This may be just as well, since major structural reorganizations are costly, create uncertainty and, if carried out too frequently, lead to attempts by lower level staff to preserve the status quo. The historical record is full of reorganizations which failed because they focused only on structure and did not address the other issues discussed in this paper.

For these reasons, structural reorganization should usually be a last-resort option. This does not mean that institutional structures are irrelevant. Structural differences have strong implications for linkage mechanisms and for the performance of IATS. We will now discuss these implications.

Interdependence. This can take the form of task interdependence (joint activities and interchanges necessary to perform a specific task) and/or resource interdependence (where one IATS component depends on another for the resources needed to perform its activities and meet its goals).

The literature has identified various types of task interdependence (Thompson, 1967; van den Val and Delbeck, 1976). The four broad categories are:

- pooled, in which each part uses a common resource base and makes a contribution to a common overall goal but there is minimal interaction between them
- sequential, where resources flow from one part to another asymmetrically
- reciprocal, when each part produces a product which is an input for the other
- team, when resources and products flow freely between all members of a communications network which combines two or more parts

Perceptions about the interdependencies between research and technology transfer activities have changed. "Initially, both activities were considered to be independent of each other but contributing to a common purpose (as in pooled interdependence). Later, extension was considered to be sequentially linked to research, receiving its inputs from research and incorporating these into a package of services for the farmer. Subsequently, their reciprocal interdependence was recognized, with extension identifying problems and supplying information which enabled researchers to define priorities" (Martínez, 1989). Finally, as in team interdependence, there is now less distinction between the two groups, with extension agents participating in experimentation and researchers coming closer to producers.

This change of concept in the literature has not, however, been fully accepted by the relevant institutions themselves. A major current linkage problem is that while research institutions tend to recognize their dependence on extension for promoting the application of research results, extension institutions frequently feel less dependent on research. In a recent survey of extension directors from 59 low-income countries, technology and linkage problems consistently received low rankings on the directors' lists of major concerns (Sigman and Swanson, 1985). Either extension directors believe sufficient technology already exists for their institutions to extend, or they give lower priority to promoting new technologies than to input distribution, credit supervision or other non-technological activities.

Given the tendency for formal boundaries to obstruct the free flow of information and other resources, in theory it might be desirable to try to organize structures so that all the people dependent on each other are grouped together in a single institution. In practice, however, this is rarely possible. Firstly, there are too many different interdependencies (Mintzberg,

1979). Secondly, factors other than interdependence must be taken into consideration when designing the structure of an IATS.

A classic example of the problems of trying to accommodate too many interdependencies through structural means can be seen where input distribution, credit supervision and the dissemination of technical information have been combined in a single agency. This improves coordination between the three activities, but dilutes the technical information component of the resulting organization to such an extent that interaction with research is sharply reduced.

The opposite can also occur. Strong research-extension links may be achieved by removing activities other than the dissemination of technical information from extension's mandate, but this will probably hinder the integration of input distribution and credit supervision. This has frequently occurred in the case of the T and V system.

Other important determinants of structural design. Besides interdependence, there are five other factors of importance in designing organizational structures. These are:

- the compatibility of the management styles required by different tasks/activities
- whether the tasks/activities involved have the same sources of legitimacy
- size considerations
- the proven capacity of different units
- differences in staff orientation

If two activities require different management styles and practices, they are generally better placed in separate units. The same holds true if they receive their political support from widely divergent groups. Administrative and supervisory economies or diseconomies of scale for different activities imply that institutions and units have a certain optimal size. There are sound arguments for assigning essential activities to a unit with a proven capacity to get the job done, even if it is not the one whose overall mandate would normally cover it. Differences in orientation among staff are another potential reason for division.

The institutional merger of research and technology transfer. Merging research and technology transfer institutions is frequently recommended in the literature as a way of increasing integration (Samy, 1986). However, bringing the two activities together in one institution is usually problematic.

In practice, research and technology transfer often exhibit surprisingly few interdependencies. Their management requirements and political constituencies are frequently divergent and somewhat incompatible. The combined institution's resulting size may be unmanageably large. The potential benefit of increased interaction may be limited by putting the two in separate units within the same institution, and the loss of autonomy caused by being in the same institution can lead to conflicts and growing resistance among personnel who see their independence increasingly threatened (Klauss, 1979).

The only situation in which bringing research and technology transfer activities together within a single institution is successful is where a system is organized around a specific region, commodity or problem. The interdependencies between research and technology transfer in these situations are much greater because both activities focus exclusively on the same crop or on the same client group. In addition, the combined size of the research and technology transfer institution is generally more manageable than it would be if broader mandates were involved.

Even if research and technology transfer are combined in the same formal organization, this will not, in itself, guarantee adequate functional links between the two activities.

Functional and market-based organizations. Another common structural issue is whether to organize the IATS on a functional basis (for example, research, extension, input distribution) or a market basis (for example, client, output, place). The evidence suggests that market-based grouping is generally more successful according to all our evaluation criteria, at least when task complexity is not very great.

Structural divisions based on function lack a built-in mechanism for coordinating the work flow. In contrast, "market-based grouping is used to set up relatively self-contained units to deal with particular work flows. Ideally, these units contain all the important sequential and reciprocal interdependencies.... And because each unit performs all the functions for a given set of products, services, clients or places, it tends to identify directly with them, and its performance can easily be measured in these terms. So markets, not processes, get the employees' undivided attention" (Mintzberg, 1979).

The empirical evidence provides qualified support for these conclusions. One study concluded that "commodity-specific extension agencies exhibited greater coordination and less conflict than did general extension agencies" (Kang, 1984). Another study found a commodity extension program performed better than general extension according to seven out of eight criteria, including the "organization of joint programs with staff of other agencies" (Ekpere, 1973).

Commodity-specific agencies may be more integrated but their integration is still far from ideal (Kang, 1984). Moreover, performance differences are sometimes more related to commodity-specific agencies' greater access to resources than to their organizational characteristics (Ekpere, 1973).

When geared towards cash crops, such agencies are relatively easy to set up and operate. It is more difficult to create them for subsistence crops and in low-resource areas, where they have problems dealing with the interactions between their crops and other elements of the farming system.

Missing tasks. Often, no unit is assigned to, or effectively carries out, one or more of the tasks necessary for the development and delivery of new technologies. Who should take on missing tasks is a difficult problem for IATS leaders.

Such tasks can be assigned either to units which already exist or to new ones. The existing units have established work patterns which would have to be altered to accommodate a new task. Hence, this task may not receive sufficient attention; or, if it does, the personnel assigned to traditional unit tasks may become resentful. On the other hand, assigning the task to a new unit inevitably creates an additional set of barriers which have to be overcome before the task can be effectively integrated with others with which it is interdependent.

To achieve high performance, there must be at least one unit responsible for, and with the capacity to carry out, the following tasks: exploratory development, technology consolidation, technology production and technology delivery—as well as to provide the links between them.

Often, it is not clear whether these tasks should be carried out by researchers or technology transfer workers, or both. Unless each group's responsibilities are clearly defined, researchers will generally prefer the task of exploratory development, while technology transfer workers will prefer the task of technology delivery. This leaves no-one to assume responsibility for technology consolidation or (to a lesser extent) technology production. McDermott calls this the "fatal gap" and argues that, unless it is filled, the division between research and technology transfer will be too wide to bridge by establishing linkage mechanisms (McDermott, 1987).

Where high performance does take place it is generally in technology consolidation and technology production that the greatest degree of integration occurs. Some linkage-related activities within these tasks are often weakly performed. These are:

- publishing and synthesizing research results
- assessing the economic and social viability of new technologies
- transforming experimental results into specific recommendations
- producing information materials for technology transfer workers
- organizing information to make past research results more accessible
- producing and distributing physical inputs

Duplication of efforts. While there are some tasks or activities for which no-one takes responsibility, there may be others in which more than one unit is involved. These are either joint activities or represent a duplication of efforts. Here, only the latter situation is discussed.

Redundancy results either from attempts to seek greater autonomy or from competition for resources. It leads to conflict between the redundant units, but is often associated with higher performance.

One reason for duplicating efforts is to increase a unit's autonomy. Rather than relying on someone else to provide information or get something done, a unit decides that it will carry out this task itself. A unit is more likely to seek autonomy if relations between it and the other unit are already strained, if it perceives the costs of the necessary coordination to be high, or if it has doubts about the capacity or motivation of the other unit to fulfil its responsibilities.

The second major reason for the duplication of efforts is competition for resources. Units take on new activities which they perceive as being of interest to donors or policy makers if this will bring them additional funding, power or prestige. In so doing they may weaken their mandate focus. The pursuit of the same activities by several units brings them into competition and often precipitates conflicts.

The existing literature is divided about whether the net result of duplication of efforts is positive or negative (Landau, 1969; Leonard, 1982b). Although the waste of resources created by duplication of efforts is frequently deplored, the worst consequence of such duplication is probably the deterioration of relations between institutions, a deterioration which results in an unwillingness to share information, learn from each other's experience and coordinate activities. On the other hand, redundancy does increase the chances of getting the job done. It permits multiple approaches to a problem, and can promote healthy competition.

The Differences between Researchers and Technology Transfer Workers

For high performance, specific linkage mechanisms are required to manage the conflicts and communication problems caused by differences between researchers and technology workers in background, training, experience, responsibilities, status and physical location.

Informal groups. Informal groups, which may or may not reflect formal divisions, have shared languages, values and attitudes, making internal communication and collaboration easier. However, as in the case of formal boundaries, such groups also lead to intergroup differences, resulting in a "them-and-us" attitude that makes intergroup communication difficult. Among the most important determinants of informal groups are differences in staff background, training, experience, responsibilities, status and physical location. Important staff background attributes include age, gender, rural or urban origin, ethnicity, nationality and educational level.

These differences have major implications for communication between researchers and technology transfer workers. One of communication research's most consistent findings is that people communicate most frequently and effectively with those who are most similar to themselves (Röling, 1989). Thus, sharp differences between research and technology transfer staff with respect to their backgrounds and other characteristics may make it very difficult for the two groups to communicate with each other.

Two particularly important differences between the two groups are their distinct work environments and responsibilities. These differences lead to different orientations with respect to goals, use of time, interpersonal relations and formality (Lawrence and Lorsch, 1967). Researchers' goals are said to be broader, less precise, but more measurable. Researchers look mostly to the broad research community for approval, whereas technology transfer workers tend to seek approval within their specific institutions. Researchers' time perspectives are supposedly longer. They are also more used to working in informal and collegial environments (Bennell, 1989).

Occupational groups. Occupational groups, such as researchers or extension agents, have many of the same characteristics as informal groups, as well as some important additional ones. They compete with each other for status and rewards. The main form this competition takes is the attempt to exclude rival groups. "Barriers to entry" are erected mainly on the basis of academic qualifications (Bennell, 1989). Thus, to justify their own status and rewards, researchers may perceive a need to distance themselves from lower status occupations such as extension.

In most low-income countries, at least outside Latin America, extension is not regarded as a professional occupation. It has also had a low status because of its association with farmers and rural life, which themselves have very low status. Generally speaking, the status distinctions between professional and subprofessional occupations are greater in developing than in developed countries, and researchers often adopt patronizing attitudes towards extension agents (Bennell, 1989). Low pay means extension services are unable to attract quality recruits, and this has only worsened the status problem.

Strong status differences between occupational groups are difficult to bridge through linkage mechanisms. These will be more difficult to design in such a way as to allow the flow of information from lower to higher status members. Where low status members have significant information about environments and technologies not well understood by researchers, poor performance will result.

In recent years extension agents have tried to solve the status problem by making their occupation more professional. This has involved taking over some activities previously performed by research, such as carrying out field trials or deciding whether to recommend a new technology, a move which has elicited mixed responses from researchers. In some cases they have resisted what they perceive as an incursion into their domain. In others, they have willingly relinquished activities to extension, but only after downgrading them and reserving the higher status activities for themselves. On rare occasions, researchers have chosen to accept an equal role with extension, and to collaborate fully.

Although differences between the two groups is a problem, so also would be too great a similarity between them. Similarity between groups erodes the unique contribution each group can make and the advantages of specialization. This implies that there is an optimum level of dissimilarity.

Personnel and Financial Management

Personnel and financial management policies and practices which encourage integration and provide flexibility in IATS result in higher levels of performance.

Differences in policies and practices between research and technology transfer institutions can greatly hinder the integration of the two activities. Policies and practices are among the contextual factors most subject to control by managers.

Recruitment, job responsibilities and training. To achieve high performance, staff should be recruited who are capable both of fulfilling their specialized tasks and of interacting effectively with other specialists. Job descriptions (as well as informal expectations) should specify the linkage-related activities required. Managers of each unit should ensure all parties involved are clear about these responsibilities.

Status, as well as links, can be enhanced by building an emphasis on collaboration into the work programs of both researchers and technology transfer staff. When a researcher is assigned to an adaptive trial run by an extension worker, this gives status and incentives to the latter. An extension worker who provides diagnostic information for developing research projects and thereby improves the design and relevance of the project improves both his/her status and that of the researcher.

In practice, these goals are rarely met in full. The pool of candidates for research and technology transfer jobs is limited and may not include people with the right qualifications and characteristics. It is hard to attract staff to some geographical areas. Communication problems may prove unsurmountable. Job descriptions are often vague, non-existent or soon forgotten. Normally, little emphasis is given to collaborative activities. High levels of graduate unemployment create pressures to hire large numbers of staff who cannot be effectively used. All this hampers an institution's ability to develop effective relations with other groups.

Limitations on the staff recruitment side can be overcome to some extent by subsequent training or work experience. To promote effective links it may be necessary to teach people additional technical or communications skills. Staff exchanges and rotations can improve knowledge of counterparts' activities and build empathy. A common orientation program or joint participation in training activities can also help to create mutual understanding. Although specialization is not abandoned, professionals in integrated IATS which regularly make relevant new technologies available usually participate in or have enough experience of the work of their technology transfer counterparts to understand and wish to enhance what the other group does.

Again, in practice, IATS often fall short of these ideals. Many training programs fail to encourage researcher-technology transfer worker interaction, provide few tools for effective interaction, and reinforce status distinctions.

Compensation. The earlier discussion of political factors pointed to the fundamental importance of incentives, at both the institutional and individual level, in promoting performance.

The most direct and effective incentives are those accruing to staff as compensation. Compensation includes salaries, honorariums, promotion opportunities, working conditions, prestige and positive feedback, fringe benefits, the attractiveness of the work involved, and opportunities for earning supplementary incomes. These benefits can be distributed on a number of different bases, one of which is the performance appraisal/evaluation of staff members. The criteria used for performance appraisal communicate the values of an organization. The emphasis given to collaboration and the types of behavior evaluated will determine the value given to linkage behavior.

Compensation affects performance in various ways. Workers perceive the rewards or punishments resulting from their performance (including their interactions with others). The levels of conflict, competition and coordination vary as a result. Compensation packages can be perceived as fair or unfair and can diminish or increase the distinctions and divisions between groups. Compensation levels and criteria which result in high levels of staff attrition and transfer can hinder effective institutional links because the parties involved have less time to develop stable expectations and communications channels.

Service orientation. No matter how enlightened the management, researchers and technology transfer workers almost always experience some tension between their duty to respond to the concerns of management and their obligation to respond to the needs of the population served. IATS in which field staff respond exclusively to management desires are rarely very successful. They also tend to have poor flows of information up the organizational chain. However, if staff respond only to demands from below, this is likely to hinder the institutions' capacity to serve as instruments of policy. The IATS with the highest performance are those in which management promotes a service orientation and allows staff sufficient flexibility to provide it, yet maintains firm control over general policy.

Financial management. The principal aspects of financial management which affect integration and performance are the sufficiency, flexibility and reliability of funding, and the existence of slack resources. Here we are referring to funding both for the IATS in general and for the financing of linkage mechanisms in particular. With respect to the latter, many IATS have practically no funding available for such key linkage-related activities as the publication of research results, visits by researchers to extension field offices, and in-house training events.

The aim of providing slack resources is to assign more resources to an activity than are strictly expected to be necessary, so as to increase the probability that the job will be completed. In our context this could mean financing redundant linkage mechanisms so as to ensure greater integration.

Integration

The role of higher authority. Often, collaboration between separate units of the IATS is ordered by a higher authority, such as a common director, an official mandate, a government regulation or plan, or a donor agency.

Instructions to collaborate usually work only when the higher body simultaneously intervenes to convince the staff concerned of the need for integration. Otherwise, the higher body must have both adequate power and sufficient information to impose its will. This is rarely the case.

The development and delivery of new agricultural technologies is a complex process and difficult to monitor closely. Instructions from above are usually vague and it is implicitly understood that not all of them can be carried out. Again, a great deal of information is lost or deliberately withheld or distorted as it moves up the hierarchy. Senior managers are beset by a wide variety of problems besides their concern for integration. In practice, research and technology transfer managers and staff have effective veto power over external efforts to achieve integration, and thus must be persuaded or motivated, as well as directed.

Failure to persuade frequently results in the creation of formal (relatively ineffective) linkage mechanisms whose principal purpose is to please superiors. In these cases open conflicts may be eliminated, but only to be replaced by more subtle forms of mutual avoidance and hostility.

Policy makers and managers can facilitate integration through the creation of superordinate goals and/or the promotion of a shared institutional culture.

Superordinate goals are those that have "a compelling appeal for members of each group, but one that neither group can achieve without the participation of the other" (Bennell, 1989). Bennell adds that such superordinate goals are only likely to be accepted when:

- the status and/or reward grievances of disadvantaged and dissatisfied groups within the IATS are adequately resolved
- individual goals are sufficiently compatible with superordinate goals
- sufficient weight is given to staff interactions in performance appraisal and rewards systems

Organizational cultures conducive to integration are easier to promote under conditions of staff homogeneity and organizational stability, and when staff have had long and intense shared experiences.

Preconditions for voluntary linkage. *Significant integration will occur only if the parties involved perceive all of the following to exist: (1) interdependence, (2) domain consensus, (3) ideological consensus, (4) domain correspondence, (5) competence and (6) the capacity to deliver on agreements.*

Since cooperation implies a certain loss of autonomy, groups will normally want to cooperate only if they perceive the potential gains to outweigh this loss. One factor in the decision whether or not to cooperate will be external pressures for improved performance, but there are also a number of strictly internal organizational factors which are important.

The first of these is whether interdependence is perceived. Both parties must feel the other has something they need. The second and third factors are domain consensus and ideological consensus. Domain consensus means that the units agree about each other's appropriate role and scope. Ideological consensus means agreement regarding the nature of the tasks confronting the units and the appropriate approaches to use of resources (Benson, 1975). For domain and ideological consensus to occur, neither unit must perceive the other's role, scope and approach as potentially threatening to its own resource base.

A fourth important factor is domain correspondence. Correspondence exists when two units share a common set of clients and topics of concern. The lack of domain correspondence between research and technology transfer institutions is a common problem. Typical examples are:

- research is organized on a national basis, while technology transfer is provincial
- research units follow agro-ecological distinctions, while technology transfer follows administrative ones
- research is divided on a disciplinary basis, while technology transfer is divided by commodity or geographical area
- research focuses on a single commodity, while technology transfer has a more general focus
- research services are targeted to one client group, technology transfer services to another

Often there is a fine line between domains being closely related and therefore complementary, and their being overlapping or even identical. Yet the likely outcomes in each case are markedly different. In the first case, task interdependencies and common orientations will be greater, facilitating interaction. In the second case, competition may arise for funds.

Competence and capacity to deliver on agreements are other necessary preconditions for voluntary linkage. If one group depends on another for resources or activities the latter is unable to provide or carry out, the first group will eventually seek alternatives which eliminate that dependence (or else use the second group's incapacity as an excuse for poor performance).

Perceptions about the other group's importance, relevance, effectiveness, efficiency and reliability are as important as whether or not these attributes really exist. Beliefs about other groups are based at least in part on stereotypes and limited information, but are heavily influenced by past experiences.

Other factors, such as a group's absolute and relative age, size, power and access to resources, have also been mentioned as affecting its inclination towards voluntary linkage. Immaturity and insecurity in organizations weaken their willingness to integrate with others. Organizations are immature if they have not yet clearly defined their domain. Insecurity implies that an organization perceives its resource base to be vulnerable.

The use of liaison positions. Liaison positions or units are sometimes used as buffers to mediate between groups which must communicate with each other. They may be within one or both of the groups, or they may form a separate entity. In the latter case the idea is that if two groups differ so much that it is very difficult for them to communicate, a third group which combines features of each of the others can act as an intermediary.

The use of such positions is often suggested as a solution to the communication problems associated with people who specialize in the different stages of technology development and delivery. Since there is typically a larger gap between researchers, technology transfer workers and farmers in low-income than in high-income countries, more liaison roles are probably needed in the former. Taken to its logical extreme, however, the communications chain could become very long. The problem with having many steps in the communication process is that the clarity and content of the information communicated diminishes with each additional link in the chain.

There is also a danger that liaison positions will accentuate rather than attenuate the integration problem. If liaison staff begin to take on the attributes of a separate group — with their own interests, beliefs, attitudes, orientations and work styles — they can become an obstacle to communication rather than a facilitator. Two mechanisms which can prevent this from happening are the incorporation of liaison positions into one of the units being integrated, and the rotation of staff assigned to liaison positions.

The use of liaison positions as intermediaries may prove more problematic ultimately than the difficulties such positions were designed to overcome. Even when liaison positions exist, they do not obviate the need for direct communication between the parties being linked.

Decentralization. *Formal and informal linkage mechanisms at several administrative levels (for example, national, regional, operational) are essential for high performance. The level of integration between researchers and technology transfer workers is higher when adaptive research is decentralized and dispersed. This higher integration leads to more relevant new technologies becoming available. Moreover, decentralization and the delegation of responsibility within an IATS require well-developed linkage mechanisms at the operational level.*

If, for example, an exchange of technical information is required, it will not be sufficient to bring together managerial staff who lack familiarity with the topic concerned. Links must also be organized between the technical staff. Conversely, regional coordination committees in highly centralized IATS frequently fail because participants cannot speak authoritatively for their institutions.

Summary

The most important environmental factors affecting IATS performance and links are: external pressure, the resources provided to the IATS for servicing its clients, and the diversity of its environments. More integrated systems, which are more successful at making available relevant new technologies, generally face strong external pressures, have access to substantial resources, and focus on simple and homogenous environments.

High-resource IATS are more differentiated than low-resource ones, with more sophisticated links to which more resources are devoted. Diverse environments are associated with the need to perform complex tasks to achieve IATS objectives. These tasks require greater professionalism, decentralization and less hierarchical management.

Less important, but still significant, environmental factors include the availability of different communications channels, the development of the necessary infrastructure and traditions for farmers to make use of inputs and information produced outside their communities, the level of pre-existing knowledge about the environment and its production systems, and the dispersion and accessibility of the farming population served. Because these factors are outside the IATS, managers have relatively little control over them. They must, however, take them into account in making decisions regarding the scope of their institution's activities, its organizational structure, its working methods, and the management of its links.

High performance requires that IATS have the responsibility and capacity to undertake the activities associated with each task in the technology development and delivery process (with the possible exception of discovery), and that identifiable functional links exist between them. In practice, the most important missing tasks tend to be technology consolidation and production. Hence these must be given special attention by managers, who are often in a good position to deal with these problems.

Different links will be required for different types of technology. In particular, activities related to already established technologies require different links to activities concerned with developing and delivering new technologies. Managers can exercise considerable control over these links.

Organizational structure, personnel management and financial management strongly affect both IATS performance and links. While the managers of technology institutions have only moderate control over organizational structure and should be cautious about exercising it, they can have greater influence over personnel policies and should take maximum advantage of that influence. Their control over financial policies is limited.

Difficult personnel problems arise from the differences between researchers and technology transfer staff in background, training, experience, responsibilities, status and physical location. These problems can greatly affect performance and need to be addressed as part of efforts to increase system integration.

Successful IATS address task and resource interdependencies through a combination of organizational grouping and linkage mechanisms. Their structural arrangements take into consideration the compatibility of the management styles required by various tasks/activities, divergences in the sources of political support for different tasks/activities, size considerations, different units' proven capacity to perform, and differences in task orientation.

A market-based grouping is generally more successful at achieving integration and relevance. However, this type of arrangement is not often feasible in diverse environments served by poorly endowed IATS.

Redundancy can have negative and positive consequences. It arises when there are strong incentives for increasing unit autonomy and competing for resources. Although it wastes requires more resources, it may ensure that objectives are met.

High levels of integration are facilitated by interdependence, domain consensus, domain correspondence, ideological consensus, competence and the capacity to deliver on agreements. The creation of superordinate goals and the promotion of an institutional culture conducive to integration are also important.

Increasing system integration is not an end in itself, but it is important because IATS that perform well according to other criteria are characterized by high levels of integration. These systems have many formal and informal linkage mechanisms, at multiple administrative levels. Many have liaison positions and departments, but these complement, rather than substitute for, more direct links.

Acknowledgments

This paper synthesizes the six papers commissioned by ISNAR as part of an international project to study links between agricultural research and technology transfer, and reproduced in this book. The project's Advisory Board played an important role in conceiving the paper. All the members of the project's core group read various versions of the paper and made useful amendments and comments. Any remaining shortcomings in this paper, however, are the responsibility of the authors.

References

The papers reproduced in this publication by Bennell, Ewell, Martínez, Pray and Echeverría, Röling, and Sims and Leonard.

Arnold, H., and Feldman, D. *Organizational Behavior*. New York: McGraw-Hill, 1986.

Benson, K. "The interorganizational network as a political economy." *Administrative Science Quarterly* (June, 1975).

Chambers, R. "Bureaucratic reversals and local diversity." *IDS Bulletin 4* (1988).

Ekpere, J. "A comparative study of job performance under two approaches to agricultural extension organization in the Midwestern State of Nigeria." Ph.D thesis. Madison: University of Wisconsin, 1973.

Israel, A. *Institutional Development, Incentives to Performance.* Baltimore: Johns Hopkins University Press/World Bank, 1987.

Kang, J. "Interorganizational relations between extension agencies and other agricultural development agencies in Asian and Oceanian countries." Unpublished Ph.D thesis. Champaign-Urbana: University of Illinois, 1984.

Klauss, R. "Interorganizational relationships for project implementation." *International Development Administration (Implementation Analysis for Development Projects).* Honadle, G., and Klauss, R. (eds). New York: Praeger Press, 1979.

Lane, H., Beddows, R., and Lawrence, P. *Managing Large Research and Development Programs.* Albany, New York: State University of New York Press, 1981.

Lawrence, P., and Lorsch, J. *Organization and Environment: Managing Differentiation and Integration.* Homewood, Illinois: Irwin Press, 1967.

Landau, M. "Redundancy, rationality, and the problem of duplication and overlap." *Public Administration Review* (4, 1969).

Leonard, D. "Analyzing the organizational requirements for serving the rural poor." *Institutions of Rural Development for the Poor: Decentralization and Organizational Linkages.* Leonard, D., and Marshall, D. (eds). Research Series 49, IIS. Berkeley: University of California, 1982a.

Leonard, D. "Choosing among forms of decentralization and linkage." *Institutions of Rural Development for the Poor: Decentralization and Organizational Linkages.* Leonard, D., and Marshall, D. (eds). Research Series 49, IIS. Berkeley: University of California, 1982b.

Leonard, D. *Reaching the Peasant Farmer: Organizational Theory and Practice in Kenya.* Chicago: University of Chicago Press, 1985.

McDermott, J.K. "Making extension effective: The role of extension/research linkages." *Agricultural Extension World Wide.* Rivera W., and Schram, S. (eds). New York: Croom Helm, 1987.

Merrill-Sands, D., and McAllister, J. "Strengthening the integration of on-farm client-oriented research and experiment station research in national agricultural research systems (NARS): Lessons from nine case studies." OFCOR Comparative Study Paper No 1. The Hague: ISNAR, 1988.

Mintzberg, H. *The Structuring of Organizations.* Englewood Cliffs: Prentice Hall, 1979.

Peters, T., and Waterman, R. *In Search of Excellence, Lessons from America's Best-Run Companies.* New York: Harper and Row, 1982.

Samy, Mohamed. "Linking agricultural research and extension in developing countries." Paper presented at the Annual Meeting of the Rural Sociology Society in Salt Lake City, Utah, 1986.

Sigman, V., and Swanson, B. "Problems facing national agricultural extension in developing countries." Champaign-Urbana: University of Illinois, 1985.

Snyder, M. "A framework for analysis of agricultural research organizations and extension linkages in West Africa." Ph.D thesis. Washington D.C.: George Washington University, 1988.

Stoop, W. "NARS linkages in technology generation and technology transfer." ISNAR Working Paper No 11. The Hague: ISNAR, 1988.

Thompson, J.D. *Organizations in Action.* New York: McGraw-Hill, 1967.

Tripp, R. "Farmer participation in agricultural research: New directions or old problems?" IDS Discussion Paper 256. Brighton: IDS, 1989.

van den Val, A., and Delbeck, A. "A task contingent model of work-unit structure." *Administrative Science Quarterly* (19, 1976).

World Bank. *Agricultural Research and Extension: An Evaluation of the World Bank's Experience.* Washington D.C.: World Bank, 1985.

Index

Printed and bound by CPI Group (UK) Ltd, Croydon, CR0 4YY

23/10/2024

01778240-0005